Michel J. Gauthier (Ed.)

Gene Transfers and Environment

Proceedings of the Third European Meeting on
Bacterial Genetics and Ecology (BAGECO-3),
20–22 November 1991, Villefranche-sur-Mer, France

With 35 Figures

Springer-Verlag

Berlin Heidelberg New York
London Paris Tokyo
Hong Kong Barcelona
Budapest

Dr. MICHEL J. GAUTHIER
INSERM Unité 303, 1 Avenue Jean Lorrain, 06300 Nice, France

Published in cooperation with the new journal MICROBIAL RELEASES

Microbial Releases

VIRUSES · BACTERIA · FUNGI

Editor-in-Chief
Walter Klingmüller
Lehrstuhl für Genetik
der Universität
Universitätsstrasse 30
W-8580 Bayreuth, FRG
FAX: (0) 921-552535

Co-Editor
Joe Shaw
Botany and Microbiology
College of Sciences
and Mathematics
Auburn University,
AL 36849-5407, USA
FAX: (205) 844-1645

Editorial Board
C. Beauchamp Québec, Canada
M. J. Cho Chinju, Korea
R. Colwell College Park, Md., USA
H. K. Das New Dehli, India
S. D. Ehrlich Jouy-en-Josas, France
P. F. Entwistle Oxford, UK
M. Gauthier Nice, France
G. Hirsch Munich, Germany
A. Kerr Glen Osmond, Australia
E. L. Kline Clemson, S.C., USA
S. T. Lam Triangle Park, N.C., USA
R. T. Lamar Madison, Wis., USA
P. A. Lemke Auburn, Ala., USA
M. Mergeay Mol, Belgium
J. A. W. Morgan Cumbria, UK
P.-P. Pastoret Brussels, Belgium
F. O. Pedrosa Curitiba, Brazil
P. H. Pritchard Gulf Breeze, Fla., USA
J. L. Ramos Granada, Spain
I. A. Tikhonovivh Leningrad-Pushkin, CIS
J. T. Trevors Guelph, Ontario, Canada
P. J. Weisbeek Utrecht, The Netherlands
E. M. Wellington Coventry, UK
J. L. White Hyattsville, Md., USA

For subscription information please contact:
For North America: Springer-Verlag New York Inc., Service Center Secaucus,
44 Hartz Way, Secaucus, NJ 07094, USA
FAX (201) 348-4505

For other countries: Springer-Verlag, Heidelberger Platz 3, W-1000 Berlin 33, FRG
FAX (0) 30-8214091

ISBN 3-540-55390-8 Springer-Verlag Berlin Heidelberg New York
ISBN 0-387-55390-8 Springer-Verlag New York Berlin Heidelberg

Typesetting: Camera-ready by author
31/3145-5 4 3 2 1 0 − Printed on acid-free paper

Scope and Goals of the New Journal
MICROBIAL RELEASES

Microbial Releases is a timely subject. Bacteria, fungi and viruses have been released for many years on many occasions, and after genetic alterations more organisms will be released in the near future. such releases are done to increase soil fertility, degrade waste, ferment fodder, make ore available, etc. Release in this sense can be defined as the introduction of manipulated or non-manipulated microorganisms into the environment in either large or small quantities for a particular, well-defined purpose.

For such releases, methods are needed for the production and application of inoculants, as well as for evaluation of the effects obtained, combined with methods to assess any risks involved. Before the release of new strains for new purposes, model studies at the laboratory level and the development of monitoring methods with sufficient sensitivity are needed. In many cases effective means of biological containment are desirable and have to be actively sought.

The topic is, by nature, highly interdisciplinary with implications for microbiology, genetic engineering, environmental science, plant science, biotechnology, molecular biology, phytopathology, plant production, soil science, ecology, food science and technology, toxicology, waste degradation, biopesticides, water protection, waste water treatment, safety (in the laboratory and outdoors), oil degradation, soil decontamination, environmental hygiene, theoretical biology and medicine, modelling, environmental laws, jurisdiction for genetic engineering, patent laws, infectious diseases (man and animal), vaccination, immunology, risk assessment, biological containment and many other topics.

Probably due to this complex situation no journal has been founded so far that focuses predominantly on microbial releases. However, there is an urgent need for it, which is now being met by this venture. The new journal shall provide specialized research workers with a forum for their studies and give others a wider overview on important questions, findings and concepts in this field. An international board of editors will guarantee the quality of the journal and see that all essential issues are covered.

There is still a widespread belief among the public, promoted in part by the media in some countries, that releases, in particular microbes – genetically altered or not – is a most dangerous subject that should not be addressed at all either experimentally or verbally. In spite of this, it is hoped that our colleagues will take the new opportunity offered here to publish their data, reviews and comments, together with comments from government agencies and regulatory offices. This approach will help to promote the understanding that mystical beliefs have nothing to do with science and that the search for the truth needs action, which then leads to better insight.

Walter Klingmüller

Preface and Acknowledgments

The present Symposium on "Gene transfers and environment" has allied the Third European Meeting on BActerial Genetics and ECOlogy (BAGECO-3) with the biennial meeting of the Microbial Ecology Section of the French Society for Microbiology (SFM). BAGECO-1 was organized by Jean-Pierre Gratia in Brussels, in April 1987, dealing on medical and epidemiological aspects of gene transfers. The BAGECO-2 meeting was organized by John Fry and Martin Day, and was held at the University of Wales College of Cardiff in April 1989. That meeting opened a large discussion on modern developments of the genetics of bacteria in aquatic and terrestrial habitats. These two previous BAGECO meetings were mainly attended by microbiologists from northern European countries. At the end of BAGECO-2, our wish, when choosing Villefranche-sur-mer for the third meeting of this series, was to favour the participation of microbiologists from southern Europe. This goal has been reached since the BAGECO-3 has been attended by some 110 participants, from Spain, Italy, Greece and France besides those coming from Belgium, the United Kingdom, the Netherlands, Germany, Switzerland, Denmark, Sweden, Ireland and the USA. The meeting was made possible by partial financial and material support from the Commission of the European Communities, the French Society for Microbiology and Rhone-Poulenc-Industries (Direction Qualité-Sécurité-Environnement). BAGECO-3 was also sponsored by the World Health Organization and the French Institut National de la Santé et de la Recherche Médicale (I.N.S.E.R.M.).

The scopes of BAGECO-3 were in the line of those selected for the previous BAGECO meetings. However, several contributions dealt with gene exchanges in human and animal gut. Most of the conferences and orally presented papers are published hereafter. This book has been organized into five main sections. The first section is devoted to contributions introducing readers to some genetic approaches, useful for the study of gene transfers in natural environments. The three following sections cover aquatic, terrestrial and intestinal habitats. The last section presents some contributions dealing with genetically engineered microorganisms (GEMs).

We take this opportunity to warmly thank all the participants and especially the authors for their cooperation in preparing their "camera-ready" manuscripts. We also wish to express our gratitude to all of those who, more or less silently, helped with the organization of the meeting and contributed to its success. Most of them were members of our INSERM Unit 303, and of the Observatoire Océanologique of Villefranche-sur-Mer. Finally, we owe our personal debt of thank to Pr.Walter Klingmüller and the members of the Springer-Verlag staff who made the publication of these Proceedings possible within reasonnable time.

The BAGECO-4 meeting will be organized by J.D.Van Elsas in 1993 at Wageningen, The Netherlands.

Villefranche-sur-Mer
June 1992   M. J. Gauthier

Scientific Committee of the Meeting

René Bailly, CNRS, Lyon, France
Berbard Baleux, University of Montpellier, France
Armand Bianchi, CNRS, Marseille, France
Jean-Claude Block, University of Nancy, France
Rita R. Colwell, University of Maryland, USA
Martin J. Day, University of Wales College of Cardiff, UK
John C. Fry, University of Wales College of Cardiff, UK
Michel Gauthier, INSERM, Nice, France
Henri Heslot, INA-PG, Paris, France
Michael G. Lorenz, University of Oldenburg, Germany
Robert V. Miller, Oklahoma State University, USA
Odette Szylit, INRA, Jouy-en-Josas, France

Secretary and Administration

INSERM Unit 303, Nice
Michel Gauthier (Organizer)
Laurence Bonneval (Administration - Secretary)
Hélène Olagnéro (Scientific Secretary)
Bernard Chabanne (Scientific Secretary)

Société Française de Microbiologie, Paris
Marie-Claire Blanchard de Vaucouleurs (Secretary)
Louis Bobichon (Treasurer)
Alain Sabatier (FSM Publications)

Contents

Section 1

TECHNICAL APPROACHES OF GENE TRANSFER IN NATURAL ENVIRONMENTS

Direct Detection of Particular DNA Sequences in Soil

S. Selenska, S. Schentzinger and W. Klingmüller
University of Bayreuth
Genetics Department
Universitätsstrasse 30
W-8580, Bayreuth
Germany

Introduction

Amplification of DNA by polymerase chain reaction (PCR) is most promising technique for monitoring of the genetic material of bacteria released into the environment. When well optimized the technique allows detection of cells that are present in low densities in the natural samples (Bej et al. 1990 ; Pillai et al. 1991 ; Brauns et al. 1991).

In our present work we have been able to amplify by PCR particular sequences of two genetically manipulated nitrogen fixing strains of *Enterobacter agglomerans* in DNA extracted from soil inoculated with 10^9 cells/g soil of each strain. The target sequences were detected at a time when conventional plating of soil samples no longer gave colonies of the studied bacteria.

Materials and methods

Bacterial strains and soil inoculations
Two genetically manipulated strains of *E. agglomerans* were used : *E. agglomerans* 19-1-1 Nal^r, Km^r/Nm^r, containing one copy of Tn5 inserted in its *nif* (low copy number) plasmid pEA9 (Klingmüller et al. 1990; Klingmüller 1991) and *E. agglomerans* D5, Rif^r, Km^r,

carrying a 1.2 kb *Pst*I fragment of pUK4, coding the APH gene of Tn*903*, on a pBR322 vector (this work). Samples of 100 g sandy loam soil from the experimental field of Bayreuth University were distributed into Erlenmeyer flasks and inoculated with approximately 10^9 bacterial cells per g soil, resuspended in saline. The flasks were then incubated at 22°C.

Extraction of DNA from soil and PCR amplifications
DNA was recovered from inoculated soil samples by the direct lysis method of Selenska and Klingmüller (1991 a, b).

PCR amplifications were performed as follows : A 400 bp fragment of the APH gene of Tn903 was flanked by two primers : 5'CAGCATTCCAGGTATTAGAA3', complementary to the (-) strand and 5'CTCACCGAGGCAGTTCCAT3', complementary to the (+) strain of the target DNA. For the amplifications of a 550 bp fragment of the *npt*II gene of Tn*5,* an oligonucleotide 5'GCAGCTGTGCTCGACGTTGTC3'was used as a downstream primer and an oligonucleotide 5'CAAGAAGGCGATAGAAGGCGATG3' was used as an upstream primer. Both sequences were amplified by AmpliTaq DNA polymerase, Stoffel Fragment (Perkin Elmer Cetus) at the same reaction conditions. The volume of the PCR samples was 50 μl and they contained : 10 mM KCl, 10 mM Tris-HCl, pH 8.3, 200 μM of each dNTP, 0.4 μM of the primers flanking the corresponding sequence, 5 mM $MgCl_2$, 0.5 μg soil DNA and 5 units of the enzyme. The DNA was melted for 3 min. at 98°C and then 50 cycles : 1 min. at 98°C, 1 min. at 50°C and 1.5 min. at 72°C, were performed. After the 50[ht] cycle, the PCR amplification was extended for 15 min. at 72°C.

The resulting samples were analysed electrophoretically in 1 % Sea Plaque agarose gel. The products of PCR amplifications were visualized by dying the gel with 1 μg/ml water solution of ethidium bromide, were cut by a sterile scalpel and digested in gel with appropriate restriction enzymes. Then an additional electrophoresis of the digested samples was performed in 4 % NuSieve 3:1 agarose gel.

Results and Discussion

As shown in Fig. 1 it was possible to detect by PCR amplifications the 550 bp sequence of *npt*II gene in a DNA recovered from inoculated soil 42 days after its inoculation with the target bacteria. The 400 bp sequence of APH gene was amplified successfully in DNA extracted from soil even later - on 56th day after inoculation. It is important to notice that Km^r colonies of the inoculated *E. agglomerans* strains were no longer detected on agar plates beyond day 41 of our experiment. We suggest that the DNA sequences detected by PCR after 41 days of exposure of target bacteria in soil conditions could represent inoculated bacteria undetectable by plating as well as extracellular DNA released from dead bacteria.

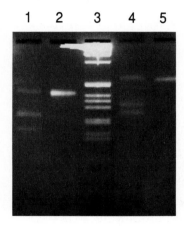

Fig.1. PCR amplifications in soil DNA samples.
1, amplification of the 550 bp fragment from the *npt* II gene in a soil DNA sample recovered 42 days after inoculation with *E. agglomerans* 19-1-1; 2, the same amplification product as in 1, but digested with *Sph* I for two hours at 37°C; 3, kilobase ladder (BRL); 4, amplification of the 400 bp fragment of the APH gene in soil DNA sample, recovered 56 days after inoculation with *E. agglomerans* D5; 5, the same amplification product as in 4, but digested with *Hind* III for 2 hours at 37°C.

It was shown that many Gram-negative microorganisms, including *Enterobacter* (Byrd et al. 1991) go into nonculturable state soon after release into the environment. Other data show that DNA released from dead organisms can stick to soil particles and so be protected from degradation (Lorenz and Wackernagel 1987). The amplification of both DNA fragments occurred at same reaction conditions in soil DNA samples. The only difference observed was that the amplification of the 550 bp fragment of the *npt*II was not successful in samples taken later than 42 days after inoculation and the amplification of the 400 bp fragment of the APH gene was successfully performed in soil samples recovered 14 days later (on the 56th day after inoculation).

One possible explanation of that difference is that in our strains the APH gene, in contrast to *npt*II gene, is localized on a high copy number plasmid. Most probably the amount of target DNA for the *npt*II gene inserted in pEA9 (a low copy *nif* plasmid of *E. agglomerans*) is below the limit of detection by PCR in soil samples taken later than 42nd day.

References

Bej AK, Steffan RJ, Di Cesare J, Haff L, Atlas RM (1990) Detection of coliform bacteria in water by polymerase chain reaction and gene probes. Appl Environ Microbiol 56:307-314

Brauns LA, Hudson MC, Oliver JD (1991) Use of the polymerase chain reaction in detection of culturable and nonculturable *Vibrio vulnificus* cells. Appl Environ Microbiol 57:2651-2655

Byrd JJ, Xu HS, Colwell RR (1991) Viable but nonculturable bacteria in drinking water. Appl Environ Microbiol 57:875-878

Klingmüller W (1991) Plasmid transfer in natural soil : a case by case study with nitrogen-fixing *Enterobacter*. FEMS Microbiol Ecol 85:107-116

Klingmüller W, Dally A, Fentner C, Steinlein M (1990) Plasmid transfer between soil bacteria. In: Fry JC, Day MJ (eds) Bacterial genetics in natural environments. Chapman and Hall, London, pp. 133-151

Lorenz M, Wackernagel W (1987) Adsorption of DNA to sand and variable degradation rates of adsorbed DNA. Appl Environ Microbiol 53:2948-2952

Pillai SD, Josephson KL, Bailey RL, Gerba CP (1991) Rapid method for processing soil samples for polymerase chain reaction amplification of specific gene sequences. Appl Environ Microbiol 57:2283-2286

Selenska S, Klingmüller W (1991a) DNA recovery and direct detection of Tn5 sequences from soil. Lett Appl Microbiol 13:21-24

Selenska S, Klingmüller W (1991b) Direct detection of nif gene sequences of Enterobacter agglomerans in soil. FEMS Microbiol Lett 80:243-246

Autotrophic Ammonia-Oxidizing Bacteria in Stratified Eutrophic Lakes: Development of Ribosomal RNA-Directed Oligonucleotide Probes

W.D. Hiorns[1], I.M. Head[1], J.R. Saunders[1], A.J. McCarthy[1], G.H. Hall[1], R.W. Pickup[1] and T.M. Embley[2]
Department of Genetics and Microbiology
University of Liverpool
P.O. Box 147
Liverpool L69 3BX
UK

Introduction

Ribosomal RNA probing has been used widely for rapid identification of clinical and environmental isolates. Novel strains can be taxonomically identified by sequence alignment and comparison. Also of interest to ecologists is the retrieval of ribosomal nucleic acids and encoding DNA directly from the environment. It has already been demonstrated that such sequences are disparate from those found in culturable microorganisms (Olsen 1990), and the importance of uncultured organisms *in situ* is thus addressed.

The oxidation of ammonia to nitrate via nitrite influences agricultural and wastewater treatment practices, yet the microbiological and ecological background of these transformations is incompletely understood (Prosser 1986). The major obstacle to bacteriological study of nitrification is the strict autotrophy of the organisms concerned; isolation of nitrifying bacteria requires

[1] Institute of Freshwater Ecology, Windermere Laboratory, Cumbria, LA22 OLP, UK
[2] Department of Zoology, The Natural History Museum, London, SW7 5BD, UK

months of enrichment and subculturing. They are also difficult to enumerate, and, importantly, the efficiency of the most commonly used method, most probable number analysis, is affected by choice of culture medium. Thus, the *in situ* significance of culturable nitrifiers is unknown.

The five genera of ammonia-oxidising bacteria contain many unnamed species, according to measurements of % G+C and DNA/DNA hybridisation (Koops and Harms 1985). Five strains of nitrosobacteria have previously been characterized by 16S rRNA oligonucleotide cataloguing (Woese et al. 1984, 1985). The representatives of *Nitrosolobus*, *Nitrosospira* and *Nitrosovibrio* are a closely related subset of the beta subdivision of Proteobacteria. A diagnostic oligonucleotide common to these genera may therefore exist. *Nitrosomonas europaea* and *Nitrosococcus mobilis* are members of a larger subset of the beta subdivision, whilst *Nitrosococcus oceanus* is of the gamma subdivision.

For preliminary ecological studies, a natural system of intense nitrification was required. The oxycline of a thermally stratified lake was chosen as a suitable site; it is aquatic, easily defined in the field and can contain up to 105-106 nitrosobacteria per litre. Large volumes were filtered by tangential flow filtration (TFF). Batch enrichment cultures were also inoculated with liquid and sediment samples which were characterised by MPN.

Materials and Methods

Strains
Nitrosomonas europaea C-31 (ATCC 25978); *N. eutropha* C-91 (ATCC 25984); *N. marina* C-56 (ATCC 25983); *Nitrosolobus multiformis* C-71 (ATCC 25196); *Nitrosovibrio tenuis* Nv-1, Nv-12, #141; *Nitrosospira briensis* C-128; *Nitrosococcus mobilis* Nc-2; *N. oceanus* C-27, C-107 (ATCC 25196). Growth was on media specified by S. Watson (personal communication). Nitrite-oxidising organisms were supplied by NCIMB and grown on recommended media.

Sequencing
Initially, reverse transcriptase sequencing of ribosomal RNA after Lane *et al.*, 1985. Linear polymerase chain reaction sequencing was later adopted (Embley 1991).

Cell blotting
Aliquots of culture were slot-blotted on to nylon membrane. Lysis was performed in the assembled manifold.

Hybridization
Oligonucleotides end-labelled with ^{32}P-ATP were hybridized according to instructions for Boehringer Mannheim's nonradioactive DNA labelling and detection kit. Hybridization was performed at the irreversible melting temperature of each probe.

Environmental sampling.
The lacustrine water column was plumbed with a combined temperature/oxygen probe, and approximately 120 litres of sample were removed from the desired level. Tangential flow filtration of this sample yielded 2 litres of concentrate which was centrifuged to recover microorganisms. Standard DNA preparation methods were then employed.

Results

Nucleic acid sequences and probes.
Complete (1.4 kilobase) sequences have been obtained for strains C-31, C-91 and C-71. Approximately 500 bases of sequence have been obtained for the eight remaining strains. Exploratory studies have been performed using oligonucleotides designed from this incomplete database, and an hierarchy of specific detections has been demonstrated. Oligonucleotide pD' (Edwards et al. 1988) was used as a positive control for eubacterial 16S ribosomal nucleic acids (oligonucleotides bearing the "prime" suffix are RNA-binding).

Oligonucleotide NMS 215'. Designed to detect a sequence from C-31 and C-91, this probe does not hybridize significantly to other nitrosobacteria, including C-56, *Nitrosomonas marina*.

Oligonucleotide AAO 258'. A region of identity was noted between rRNA sequences from C-31, C-91 and C-71. This probe hybridizes to nucleic acids from a broader range of nitrosobacteria, including members of *Nitrosospira* and *Nitrosovibrio* (Fig. 1).

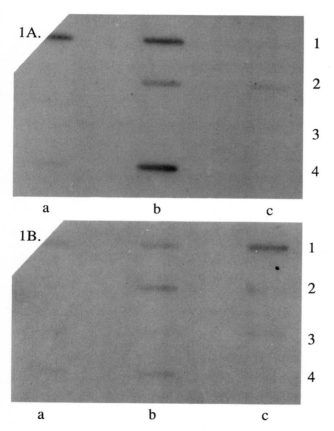

Fig. 1. Cell blots of ammonia-oxidizing bacteria probed with AAO 258' (1A) and pD' (1B). Strains are: 1a, *Nitrosomonas europaea* C-31; 1b, *N. eutropha* C-91; 1c, *N. marina* C-56; 2a, *Nitrosovibrio tenuis* Nv-1; 2b, *N. tenuis* Nv-12; 2c, *N. tenuis* #141; 3a, *Nitrosococcus oceanus* C-27; 3b, *N. oceanus* C-107; 3c, *N. mobilis* Nc-2; 4a, *Nitrosospira briensis* C-128; 4b, *Nitrosolobus multiformis* C-71.

Environmental applications.

Nucleic acids isolated from lakewater were probed with a range of diagnostic oligonucleotides. Nucleic acids were also recovered from actively nitrifying enrichments seeded with lakewater and sediment dilutions. Although rRNA sequences were detected, no significant hybridization was noted between any combination of sample and diagnostic probe (Fig 2). Importantly, the density of the native ammonia-oxidizing population is estimated by viable enumeration to have been 1-10 cells per litre, a consequence of high winds having disturbed the water column prior to sampling.

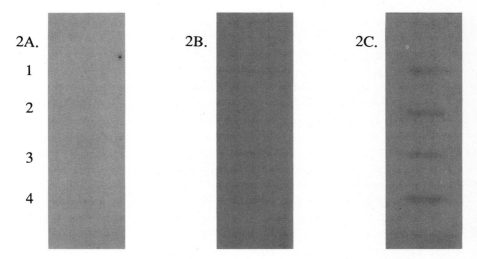

Fig. 2. Purified DNA probed with NMS 215' (2A), AAO 258' (2B) and pD' (2C). Samples are: 1, TFF enrichment; 2, sediment enrichment; 3, sediment MPN (10^{-7}); 4, sediment MPN (10^{-8}).

Discussion

The major obstacle to this work has been the acquisition of a representative set of ammonia-oxidizing bacteria. This has largely been overcome with the assistance of Professor Stanley Watson (Woods Hole Oceanographic Institution, Massachusetts), who has contributed the eight incompletely-characterized strains. It is anticipated that AAO 258' will recognize many closely related members of the nitrosobacteria, especially those of the genera *Nitrosolobus*, *Nitrosospira* and *Nitrosovibrio*.

The failure to detect native lacustrine DNA with "nitrifier" probes is unsurprising, the nitrifying community apparently being dispersed. However, enrichment cultures containing very active communities also evade detection. We therefore wish to clone and sequence ribosomal RNA genes from enrichments and native populations. Generic "nitrosobacterial" oligonucleotides are a foundation for such work. In this way, our tentative findings could be substantiated. Nevertheless, the potential of this analytical approach has been demonstrated.

References

Edwards U, Rogall T, Blocker H, Emde M, Bottger EC (1989) Isolation and direct complete nucleotide determination of entire genes. Characterisation of a gene coding for 16S ribosomal RNAs. Nucleic Acids Res 17:7843-7853.

Embley TM (1991) The linear PCR reaction: a simple and robust method for sequencing amplified rRNA genes. Lett Appl Microbiol 13:171-174.

Koops H-P, Harms H (1985) Deoxyribonucleic acid homologies among 96 strains of ammonia-oxidising bacteria. Arch Microbiol 141:214-218.

Lane DJ, Pace B, Holsen GJ, Stahl DA, Sogin ML, Pace NR (1985) Rapid determination of 16S ribosomal RNA sequences for phylogenetic analyses. Proc Nat Acad Sci 82: 6955-6959.

Olsen GJ (1990) Variation among the masses. Nature 345:20.

Prosser JI (1986) Preface. In JI Prosser (ed), *Nitrification*. IRL Press, Oxford, Washington DC, pp. v.

Woese CR, Weisburg WC, Hahn CM, Paster BJ, Zablen LB, Lewis BJ, Macke TJ, Ludwig W, Stackebrandt E (1984) The phylogeny of purple bacteria: the beta subdivision. Syst Appl Microbiol 5:327-336.

Woese CR, Weisburg WC, Hahn CM, Paster BJ, Zablen LB, Lewis BJ, Macke TJ, Ludwig W, Stackebrandt E (1985) The phylogeny of purple bacteria: the gamma subdivision. Syst Appl Microbiol 6:25-33.

The Detection of Marked Populations of Recombinant Pseudomonads Released into Lake Water

C. Winstanley, J.A.W. Morgan[1], J.R. Saunders and R.W. Pickup[1].
Department of Genetics and Microbiology,
University of Liverpool,
Liverpool L69 3BX
UK

Introduction

Some of the more esoteric catabolic activities exhibited by bacteria in their battle to degrade recalcitrant compounds, many of which are man-made pollutants, have been observed and studied extensively in species of *Pseudomonas*. Much research has been concentrated on elucidating the mechanisms of the catabolic pathways involved, with the result that many of the genes, often plasmid-borne, have been cloned and their regulatory systems resolved. The advance of recombinant DNA technology has allowed the manipulation of such catabolic activities to improve efficiency or range. This has encouraged the suggestion that genetically-modified microorganisms may be employed in the environment to degrade pollutants *in situ*. The success of such operations depends largely on the ability of laboratory-manipulated strains to survive and function following release into an unwelcoming environment.

---

[1] Institute of Freshwater Ecology, Ambleside, Cumbria LA22 OLP, UK

xylE marker systems

In order to monitor the survival of populations of recombinant *Pseudomonas putida* released into lake water model systems, we have developed techniques to allow the genetic tagging of target microorganisms, and methods for subsequent detection. Our chosen marker gene, *xylE* (catechol 2,3 oxygenase; C23O), has been placed under the control of strong bacteriophage lambda promoters, pL or pR, with the gene $\underline{cI}$857, encoding a temperature-sensitive repressor protein, included in marker cassettes when regulation of expression is desired. This enables us to study any deleterious effects caused by very high expression of the marker gene. Three marker cassettes were constructed: $\underline{P}$L-*xylE*; $\underline{P}$L-*xylE*-$\underline{cI}$857; $\underline{P}$R-*xylE*-$\underline{cI}$857; all of which were introduced into the broad-host-range IncQ plasmid pKT230 to enable study of the stability, expression and regulation of *xylE* in a range of Gram-negative hosts. The $\underline{P}$L-*xylE*-cI857 cassette was poorly regulated in some hosts, including pseudomonads. The $\underline{P}$L-*xylE* (unregulated) and $\underline{P}$R-*xylE*-$\underline{cI}$857 (regulated) systems functioned well in all the hosts tried (Winstanley et al. 1989) and were subsequently introduced into the IncP conjugative plasmid R68.45 (Winstanley et al. 1991). Marker cassettes have also been introduced directly into the chromosome of *P. putida* by means of transposon delivery systems. A summary of the marker systems employed is presented in Table 1.

Table 1. *xylE* marker systems

a. Plasmid-located

Plasmid	Inc Group	Genetic character[1]	Phenotype	Size[2]
pLV1010	Q	pL-*xylE* Sm^R Ap^R	unregulated	17.1
pLV1011	Q	pR-*xylE*-$\underline{c}$l857Sm^R	regulated	16.5
pLV1013	Q	pR-*xyl*-$\underline{c}$l857Sm^{R}Km^R	regulated	14.2
pLV1016	P	pR-*xylE*-$\underline{c}$l857 Ap^{R}Km^{R}Tc^R	regulated	68.1
pLV1017	P	pL-*xylE* Ap^{R}km^{R}Tc^R	unregulated	67.3

Table 1. Continued

b. Chromosomally-located

Marker cassette	Phenotype
Tn5-pR-xylE-cl857	unstable, regulated
Tn7-pL-xylE	stable, unregulated

[1] Ap^R, KM^R, Sm^R and Tc^R and Tc^R indicate resistance to ampicillin, kanamycin, streptomycin and tetracycline respectively.
[2] Sizes are given in kilobase-pairs

Detection methods

A number of methods have been developed to detect marked populations of *P. putida* following release into lake water model systems (Morgan et al. 1989). All release experiments were performed at 10°C over a 28 day period using lake water recovered from the surface of Lake Windermere, Cumbria, United Kingdom. Both sterile-filtered and untreated lake water, in volumes of 100 ml to 1 litre, were used.

Direct enzyme assay
Cells can be recovered from lake water by filtration, resuspended in buffer, sonicated and centrifuged leaving a cell extract which can be assayed directly for C23O activity. This method was used to confirm that the enzyme was being expressed *in situ*. The limit for detection in untreated lake water was 10^3 cells ml^{-1}.

Hybridization.
Using *xylE*-specific or *xylE* mRNA-specific oligonucleotide or whole gene probes, it is possible to detect marker DNA or mRNA following extraction from filtered water samples. This method can be improved greatly by using PCR amplification. In order to gain an indication of the activity of the cells, the amplification of mRNA is desirable.

Immunological methods

Using filtered water samples, it is possible to detect 10^3 cells ml^{-1} of marked *P. putida* by ELISA employing C23O polyclonal antibodies. This method also indicates the active production of C23O *in situ*.

We have developed an immunocapture strategy for the recovery of our model host strain, *P. putida* PaW8, from environmental samples (Morgan et al. 1991). The methodology involves use of a monoclonal antibody, MLV1, which is highly specific for the flagellin of PaW8. We have cloned and sequenced the *P. putida* PaW8 flagellin gene in order to identify the antigenic region with a view to comparing the strain-specific sequences with those of other closely related *P. putida* strains.

Magnetic polystyrene beads, coated with MLV1 are used to attach target cells in water samples. Bead-cell complexes can be recovered by use of a strong magnet, and the cells washed off and quantified.

Culture methods

A simple colorimetric spray test enables us to detect colonies expressing *xylE* on non-selective media. Survival studies carried out in sterile-filtered or untreated lake water gave rise to similar results whether measured by numbers of colony forming units (CFU) or direct detection methods. No non-culturable but viable population was detected. This suggested that, for these experiments, culture methods are a valid indicator of survival.

In sterile lake water, populations of *P. putida* were able to grow to a sustainable level of 10^5-10^6 CFU ml^{-1}. The stability of, and metabolic burden imposed by, marker DNA strongly influenced the proportion of this sustainable population retaining *xylE*. Loss of marker DNA during growth was much more apparent in populations marked with unregulated marker cassettes. Release experiments carried out in untreated lake water revealed no significant difference in survival of populations of *P. putida,* as measured by detection of *xylE,* between either: hosts carrying regulated or unregulated marker cassettes; marker cassette location on IncP or IncQ marker plasmid or in the chromosome; auxotrophic or prototrophic host strains; stable (during growth) or unstable marker DNA. Populations, released between 10^2 and 10^5 CFU ml^{-1}, all underwent a gradual decline towards becoming undetectable within 28 days. Those released at lower cell densities became undetectable earlier.

A great deal of variability in the survival of populations of the same strain in different water samples was observed.

Conclusion

Before any large scale release into the environment can be considered, it is a sensible first step to study small-scale model systems. In our experiments with populations of *P. putida*, we have been able to observe differences in survival only when sterile-filtered water was used, or when the water sample itself was varied. The various methods we have employed to mark our target organism did not appear to influence survival in untreated lake water, even when the marker DNA was highly unstable during growth. Since the DNA content or stability had no apparent effect on survival, as measured by detection of *xylE,* it seems likely that populations of *P. putida* released into lake water decline to undetectable levels whilst undergoing little or no turnover of cells. Observations during experiments using the direct C23O assay and ELISA methods, however, suggest that the C23O enzyme is produced during this period of decline. The possibility exists, therefore, that a released population, even whilst undergoing steady decline, could carry out its intended function.

References

Morgan JAW, Winstanley C, Pickup RW, Jones JG, Saunders JR (1989) Direct phenotypic and genotypic detection of a recombinant pseudomonad population released into lake water. Appl Environ Microbiol 55:2537-2544

Morgan JAW, Winstanley C, Pickup RW, Saunders JR (1991) Rapid immunocapture of *Pseudomonas putida* cells from lake water by using bacterial flagella. Appl Environ Microbiol 57:503-509

Winstanley C, Morgan JAW, Pickup RW, Jones JG, Saunders JR (1989) Differential regulation of lambda pL and pR promoters by a cI repressor in a broad-host-range thermoregulated plasmid marker system. Appl Environ Microbiol 55:771-777

Winstanley C, Morgan JAW, Pickup RW, Saunders JR (1991) Use of a *xylE* marker gene to monitor survival of recombinant *Pseudomonas putida* populations in lake water by culture on nonselective media. Appl Environ Microbiol 57:1905-1913

Screening of Erythromycin Resistant *Bacillus* spp. from Aerobiological Samplings for Recombinant DNA

K. Smalla
Biologische Bundesanstalt für Land- und Forstwirtschaft
Institut für Biochemie und Pflanzenvirologie
Messeweg 11/12
D-3300, Braunschweig
Germany

Introduction

We performed a case study for an industrial alpha-amylase production with a genetically engineered *Bacillus subtilis* strain. Contrary to the original host strain, the applied genetically engineered strain carried 4 to 6 copies of an erythromycin (Em) resistance plasmid with a duplicated alpha-amylase gene insert. The used plasmid vector was a deletion derivative of the streptococcal resistance plasmid pSM19035 (Steinborn and Hofemeister 1984). Under large-scale fermentation conditions the alpha-amylase production was increased 3-4 times due to the gene dose effect. Furthermore, a plasmid-mediated Em-resistance was expressed.

Our investigations demonstrated that an unintended release of the rDNA production strain from the fermentation plant occurred (Smalla et al. 1991). The main routes of release were shown to be air (exhaust air, aerosols), biomass and waste water.

Possible adverse effects due to an unwanted release of *B. subtilis* strain into the environment were seen in an additional spread of the streptococcal Em-resistance gene. Such a hypothetic additional spread could be a result of the survival, multiplication and transport of the rDNA strain in natural ecosystems or of gene transfer events. Gene transfer might occur especially when fermentations are infected, in biomass and waste water because of very high densities of bacterial cells, for instance in biomass 10^9-10^{10} CFU/g.

The intention of the investigations presented here was :
- to collect information about the incidence of Em-resistant Gram-positive air-borne bacteria in the surroundings of the fermentation plant,
- to check Em-resistant *Bacillus* species from aerobiological samplings for rDNA.

Characterization of air-borne bacteria

During large-scale production of alpha-amylase with the rDNA *B. subtilis* extending over a period of four years, microbial air contaminations were measured by settle plates and impactors as described by Smalla et al. (1991). Air-borne bacteria were identified roughly, the proportion of erythromycin-resistant bacteria and antibiotic resistance pattern were determined and compared with investigations of air-borne microbes in the surroundings of the hygiene institute in Magdeburg, located 30 kilometers away from the fermentation plant.

About 1300 Gram-positive bacteria randomly isolated from aerobiological samplings taken from surroundings of the fermentation plant (A) and 431 from the control site (B) were identified according to Aislabi and Loutit (1984). For air-borne bacteria from A we found that 67% of the tested isolates belonged *Bacillus* spp., 9% to *Streptococcus* spp., 17.5% to *Micrococcus* spp. and 6.7 to *Staphylococcus* spp. Except for *Streptococcus* spp. where 18% of the tested isolates were Em-resistant, the proportion of Em-resistant isolates was about 5% for *Bacillus, Staphylococcus* and *Micrococcus* spp. The proportion of Em-resistant Gram-positive bacteria was similar and even lower than that found for isolates from B. Our results showed the natural occurrence of Em-resistance phenotype in bacilli. Em-resistance phenotype in *B. licheniformis* is determined by chromosomally located ermD genes as described by Gryczan et al. (1984). Contrary to that, the rDNA *B. subtilis* contains a plasmid located ermB gene (Ounissi and Courvalin 1981).

Most Gram-negative bacteria are intrinsically resistant to erythromycin. We found also a higher proportion of single and multiple resistant Gram-positive bacteria from aerobiological samplings of B compared with A. These findings might be attributed to differences in the bacterial composition of both sites (Smalla et al. 1991).

Screening of Em-resistant bacilli for r-DNA

More than 60 Em-resistant *Bacillus* species from aerobiological samplings of A were thoroughly characterized for their phenotypic and genotypic characteristics. To check whether rDNA is detectable Em-resistant *Bacillus* species were investigated by traditional plating and enrichment techniques, DNA-hybridization of colony and dot blots. The taxonomic identification of Em-resistant bacilli isolated from aerobiological sampling plates was performed by E. Juhr, Berlin, using morphological and biochemical properties as well as fluorescence antibody microscopy. Identification showed that tested bacilli belonged to *Bacillus licheniformis* (82%), *B. subtilis* (8%), *B. amyloliquefaciens* (7%) and *B. cereus* (3%).

About 55% of the analyzed Em-resistant bacilli had a multiple resistance against the tested antibiotics. Plasmid screening indicated that about 35% of Em-resistant *Bacillus* spp. harbored plasmids, 14% carried four and more plasmids. Plasmids of low molecular weight (less than 20kb) predominated.

Screening of *Bacillus* species for heavy metal tolerance showed that all tested strains were sensitive to mercury and cadmium whereas a high proportion was tolerant to arsenic, copper and chromium. All tested bacilli tolerated tellur.

The specificity of the digoxigenin labeled C-fragment for the recombinant plasmid was tested with 85 different *Bacillus* strains (23 different *Bacillus* species) from a *Bacillus* culture collection. None of the tested bacilli gave a positive signal after hybridization with the labeled C-fragment which proves the specificity of the gene probe for tracking pSB20. One of the tested 60 Em-resistent air-borne bacilli from A gave a positive signal in colony hybridization. After genomic DNA-extraction we obtained three positive signals in dot blots. This result underlines that efficient lysis of the blotted colonies might be a crucial point for DNA detection in bacilli. The three strains with DNA-homology to the C-fragment belonged to *B. licheniformis* and showed different phenotypic properties (antibiotic and heavy metal resistance patterns). Plasmid extraction and detection indicated plasmids of size comparable to the recombinant plasmid pSB20 for two of these strains. However, restriction pattern of HindIII-digests showed no correspondence to pSB20 restriction patterns. Therefore, we assume that detection of C-fragment homologous DNA in *B. licheniformis* may be attributed to natural gene exchange between streptococci and bacilli.

Conclusion

Our results demonstrated the occurrence of Em-resistance phenotype in bacilli from natural environ-ments. The investigation of air-borne microbes from A and B showed no significant differences with regard to the frequency of antibiotic resistant bacteria. The proportion of Em-resistant bacteria was even higher in the control group. We could not show that the unwanted release of *B. subtilis* with Em-resistance plasmids increased the proportion of air-borne bacteria resistant to erythromycin. The screening of Em-resistant bacilli gave no convincing evidence for natural transfer of the recombinant plasmid to indigenous bacilli.

References

Smalla K, Isemann M, John G, Weege K-H, Wendt K, Backhaus H (1991) A risk assessment of industrial production of alpha amylase with an rDNA productionstrain. in: Biological monitoring of genetically engineered plants and microbes. MacKenzie DR, Henry SC. Proceedings of the Kiawah Island conference. p. 205-220

Aislabi J, Loutit MW (1984) The effect of effluent high in chromium on marine sediment aerobic heterotrophic bacteria. Mar Environ Res13:69-79

Steinborn G, Hofemeister J (1984) Verfahren zur Herstellung von alpha amylase. WP DD 233 852 B1

Gryczan TJ, Israeli-Reches M, DelBue M, Dubnau D (1984) DNA sequence and regulation of ermD, a macrolide-lincosamide-streptogramin B resistance element from *Bacillus licheniformis*. Mol Gen Genet 194:349-356

Ounissi H, Courvalin P (1981) Classification of macrolide-lincosamide-streptogramin-B-type antibiotic resistance determinants. Ann Microbiol (Institut Pasteur) 132:441-454

Stress and Survival in *Alcaligenes eutrophus* CH34: Effects of Temperature and Genetic Rearrangements.

D. Van der Lelie, A. Sadouk[1], A. Ferhat[2], S. Taghavi, A. Toussaint[2] and M. Mergeay
Laboratory of Genetics and Biotechnology
V.I.T.O.
Boeretang 200
B-2400, Mol,
Belgium

Introduction

Alcaligenes eutrophus, a chemolithotrophic Gram-negative bacterium, is a good colonizer of polluted soils perfectly apt to survive in a variety of harsh conditions (Diels et al. 1989; Diels and Mergeay 1990). Therefore, strains of this species attracted our interest for various environmental applications. Using recombinant *A. eutrophus* CH34 (ATCC 43123) and *A. eutrophus* A5 strains, we were able to obtain concomitant expression of xenobiotic degrading genes (derived from the A5 chromosome) and heavy metal resistances encoded by the CH34 megaplasmids pMOL28 and pMOL30 (Mergeay et al. 1985; Springael Ph.D. thesis). Such strains look promising in bioremediation processes. In addition, research is going on to use CH34 or related strains in water and soil depollution and metal reclamation. *Alcaligenes eutrophus* was not only shown to be a good host for the expression of foreign genes, but also easily accessible for genetic manipulation by means of conjugaison. Having

---

[1] Present address : Institut Pasteur, 1 rue Calmette, F-59019, Lille Cedex, France.
[2] Laboratoire de Génétique, U.L.B., B-1640, Rhode-St.-Genèse, Belgium.

in mind to use *A. eutrophus* as a containment system for the release of recombinant DNA in the environment, this bacteria was also used as a recipient in a study to establish a model system assaying the intergeneric transmission and expression of cloned genes in soil samples (Top et al. 1990).

Although the metallotolerant strains of *A. eutrophus* were shown to survive under harsh conditions, various independent isolates of this bacteria displayed strange survival features when grown on rich media at 37°C (its normal growth temperature is 30°C). A high mortality was observed and the frequency of survivors varied between 10^{-5} to 10^{-3} as compared to the viable count at 30°C (Diels and Mergeay 1990). Some features of this peculiar phenomenon are described in the present paper/communication.

Results and Discussion

Analysis of survivors of strain CH34 after incubation at 37°C revealed the presence of many different mutations. A minor group of mutants suffered from substantial plasmid rearrangements. One class of such mutants in which the megaplasmids pMOL28 and pMOL30 were replaced by a new plasmid of 210 kb, designated pMOL50, exhibited the following features : loss of pMOL30 (Czc⁻ phenotype : loss of the resistances to Cd^{2+}, Co^{2+} and Zn^{2+} associated with the *czc* operon of pMOL30 (Diels et al. 1989; Nies et al. 1987; Mergeay et al. 1985) ; acquisition of a 45 kb insertion in a 13.3 kb *Eco*RI fragment of pMOL28. This insertion seems to consist of DNA from both plasmid and chromosomal origin : excision of an insertion sequence designated IS1086 from pMOL28. The rearrangements in pMOL28, leading to the formation of pMOL50, were shown to result in derepression of selftransfer. While the parental plasmids pMOL28 and pMOL30 could be transferred at low frequencies (10^{-7} transconjugants per donor cell), pMOL50 was transferred at high frequency (10^{-3} transconjugants per donor cell). In addition, pMOL50 was shown to mediate the transfer of chromosomal markers (10^{-6} to 10^{-5} transconjugants per recipient). This mobilization is accompagnied by a strong linkage between the markers, in such way that 6 matings were sufficient to achieve the circularity of the *A. eutrophus* CH34 chromosome. This feature of pMOL50 was used to obtain the chromosomal map of strain CH34.

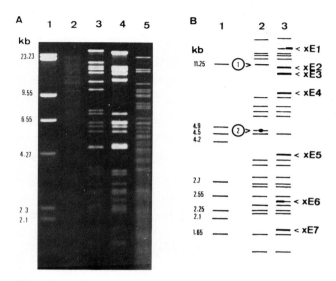

Fig. 1. A) Endonuclease *Eco*RI restriction pattern of AE104 chromosome (lane 2), pMOL50 (lane 3), pMOL28 (lane 4), pMOL30 (lane 5) and λ-*Hind*III as Wt. Marker (lane 1).
B) Schematic presentation of endonuclease *Eco*RI restriction pattern of pMOL28 (lane 2), pMOL50 (lane 3) and λ-*Pst*I as Wt. Marker (lane 1). xE1 is the left border of the insertion and xE6 the right border.
(1) bE5 (13.3 kb) : site of the 45 kb insertion leading to pMOL50.
(2) bE11 (4.8 kb) : fragment containing IS*1086*. Excision of IS*1086* results in xE5.

The mutations most frequently found among the survivors at 37°C were of chromosomal origin. Up to 5 to 80 % of these survivors exhibited recognizable mutations, like Aut⁻ (unable to grow autotrophically), Ntr⁻ (unable to use nitrate in anaerobics), Thr⁻ and Lys⁻. In some survivors, combinations of 2, 3 or even 4 phenotypes were found, indicating the presence at 37°C of a very active mutation mechanism. This temperature inducible mutation mechanism was nicknamed "Thermospontagenesis" (TSP).

To study thermospontagenesis in more detail, we introduced pJV240 that contains the sacRB operon encoding the B. subtilis levansucrase, in strain AE104, the plasmid free derivative of CH34. Expression of the sacRB operon is, in the presence of 5 % sucrose, toxic for Gram-negative bacteria (Steinmetz et al. 1983). Cultures of AE104 (pJV240), grown at 30°C, were plated on medium with and without 5 % sucrose and incubated at 30°C or 37°C, this in order to determine the percentage of sucrose resistant mutants obtained by direct or indirect selection after incubation at these two different temperatures. At 30°C, the frequency of Suc^R colonies obtained by direct selection was 10^{-4}, but at 37°C a frequency of 0.65 was observed as compared to the survivors at 37°C on plates without sucrose. From the colonies isolated after "thermospontagenesis" on plates without sucrose, 15 % were sucrose resistant, this in contrast with indirect selection at 30°C, where no sucrose resistant mutants could be found among 1000 colonies tested. The types of mutations leading to Suc^R were also examined. At both 30°C and 37°C, 10 % to 50 % of the Suc^R mutants contained a pJV240 in which the sacRB operon was inactivated by an insertion element. In addition, small or large deletions that might also have been caused by IS elements were found, together with non-characterized chromosomal mutations leading to Suc^R. This suggests that at both temperatures the same mutation mechanism is active, all be it much more (> 10.000 times) at the higher temperature.

Since thermospontagenesis seems to be correlated with a high activity of IS elements, we propose that at 37°C a general control mechanism regulating the activity of IS elements is deregulated.

Because thermospontagenesis was only observed at 37°C on rich medium and not on minimal medium we examined which medium compound would be responsible for this phenomenon. It was found that the presence of methionine or some of its precursor metabolites resulted in a mortality comparable to that observed during TSP. (See Table 1). In addition, similar effects were observed for cysteine and serine, also only at 37°C. Therefore we postulate that both thermospontagenesis and the effects observed with serine, cysteine and methionine are due to changes in the activity of a central regulatory element in *A. eutrophus* when cultivated at 37°C. This central regulatory element might also play a role in the adaptation of *A. eutrophus* towards harsh environmental conditions.

Table 1 : Effects of methionine and its biosynthesis precursors on thermospontagenesis with *A. eutrophus* CH34 and AE53.

| | Numbers of survivors | | |
Medium	30°C	37°C	37°C/30°C ratio
869 (rich)	8.0×10^9	3.3×10^6	4.1×10^{-4}
284 gluconate (minimal)	6.8×10^9	4.0×10^9	0.6
284 gluconate + methionine	7.8×10^9	1.7×10^6	2.2×10^{-4}
284 gluconate + homo-cysteine	5.2×10^9	1.8×10^7	3.5×10^{-3}
284 gluconate + S-adenosyl-methionine	7.1×10^9	1.7×10^8	2.4×10^{-2}
284 gluconate + S-adenosyl-homo-cysteine	6.6×10^9	1.3×10^9	0.2

References

Diels L, Sadouk A, Mergeay M (1989) Large plasmids governing multiple resistances to heavy metals : a genetic approach. Toxicol Environ Chem 23:79-89

Diels L, Mergeay M (1990) DNA probe-mediated detection of resistant bacteria from soils highly polluted by heavy metals. Appl Environ Microbiol 56:1485-1491

Mergeay M, Nies D, Schlegel HG, Gerits J, Charles P, Van Gijsegem F (1985) *Alcaligenes eutrophus* CH34 is a facultative chemolithotroph with plasmid-bound resistant to heavy metals. J Bacteriol 162:328-334.

Nies D, Mergeay M, Friedrich B, Schlegel HG (1987) Cloning of plasmid genes encoding resistant to cadmium, zinc and cobalt in *Alcaligenes eutrophus* CH34. J Bacteriol 169:4865-4868

Steinmetz M, Le Coq D, Djemia HB, Gay P (1983) Analyse génétique de *sac*B, gène de structure d'une enzyme secrétée, la lévane-saccharase de *Bacillus subtilis* Marburg. Mol Gen Genet 191:138-144

Top E, Mergeay M, Springael D, Verstraete W (1990) Gene escape model : transfer of heavy metal resistance genes from *Escherichia coli* to *Alcaligenes eutrophus* on agar plates and in soil samples. Appl Environ Microbiol 56(8):2471-2479.

Section 2

AQUATIC ENVIRONMENTS

Effect of Host Cell Physiology on Plasmid Transfer in River Epilithon

M.J. Day, J.C. Fry and J. Diaper
School of Pure and Applied Biology
University of Wales College of Cardiff
P.O. Box 915
Cardiff
Wales
UK

Introduction

The current interest in gene exchange between bacteria in natural populations stems from two basic and different views which overlap to some degree. Academic curiosity largely lies in attempts to achieve an understanding of when, how and what are the regulatory elements and environmental factors governing the frequency of gene transfer. Commercially driven interests provide a demand to know what are the environmental risks associated with the release of a unique recombinant, arising from the transfer, survival and expression of genes in this recombinant. The general rules derived from a few examples, in the first instance, may be of some value, but conclusions based on such a paucity of knowledge are uncertain, due to our lack of knowledge of the extent of the ecological and genetical interactions which occur naturally. This clearly illustrates the need to extend the quantity of research in this area, to cover several environments, and to examine a range of organisms. Only from such an analysis, one done in more depth and breadth, can factors of prime importance in promoting or restricting transfer be identified.

A survey of the literature clearly shows the extent of research directed at examining gene exchange between bacteria. The experiments, with few exceptions, are done in the laboratory or in simple microcosms (Fry and Day 1990). Although of value to producing an understanding of how recombinants behave in the laboratory, these studies are of unknown ecological relevance. Thus studies that compare laboratory and *in situ* results are crucial to the development of a sound understanding of the contribution gene exchange makes to the evolution of microbial populations, whether it be by the transfer of natural or recombinant genes.

Plamids are widespread amongst bacteria in nature (Kobori et al. 1984; Aznar et al. 1988; Wortman and Colwell 1988; King 1989). Their distribution varies and they tend to be more prevalent in polluted environments (Hada and Sizemore 1981; Burton et al. 1982; Baya et al. 1986; Wickham and Atlas 1988).

Transfer of plasmids to and from pure cultures of natural isolates has been shown (Shaw and Carbelli 1980; Corliss et al. 1981; Gauthier et al. 1985; Genthner et al. 1988). Studies on the influence of abiotic factors such as temperature have also been done (Walmsley 1976; Smith et al. 1978; Singleton and Anson 1981; Kelly and Reanney 1984; Rochelle et al. 1989a). The factors considered in this paper which influence plasmid transfer are temperature and restriction, together with the physiology and physiological status of the interacting cells. Previously, Reanney et al. (1982) had suggested that the physiology of the donors and recipients could also play a major role in conjugaison, but very few reports have specifically demonstrated transfer into bacteria of different physiological types (Kolenc et al. 1988). The aim of this research was to establish which factors influenced the transfer success of a natural, exogenously isolated, plasmid, pQM1 (Bale et al. 1987).

Many studies have been done in aquatic environments which have been termed *in situ*. However, the majority of these reports were based on experiments done in some form of enclosed system. The environments examined include waste water treatment plants (Mach and Grimes 1982; Gealt et al. 1985) as well as fresh water systems (Altherr and Kasweck 1982; Gowland and Slater 1984; Trevors and Oddie 1986; Cruz-Cruz et al. 1988; O'Morchoe et al. 1988). These experiments exposed the organisms to environmental factors such as pH and temperature, but with the exception of O'Morchoe et al. (1988), excluded the natural bacterial community. To assess the importance of plasmid transfer in the environment unenclosed experiments are necessary (Fry and Day 1990); but it is only recently that plasmid transfer has been demonstrated without enclosure. These experiments have been done on stones in the River Taff both with and without an intact epilithon (Bale et al. 1987; 1988a; 1988b).

Epilithon

The epilithon of freshwater rivers contains a dense population of closely spaced bacteria (Lock et al. 1984), and so is an ideal community for studying plasmid transfer in water. It is a rich and complex environment containing an assortment of bacteria, algae, protozoa and invertebrates (Lock 1981; Lock et al. 1984). It is composed of a community of organisms within a slimy layer (the biofilm) which develops on all surfaces in aquatic habitats. The combination of photosynthesis from algae and the organic products they secrete, together with renewal of nutrients, from the overlaying water, ensures maintenance of a nutrient rich environment in the biofilm. The epilithon in the River Taff forms a dense community of bacteria with viable counts about 9×10^6 bacteria. ml^{-1}, about 9% of the total count (Fry and Day 1990). The viability determined by a microcolony method (Fry and Zia 1982) is about 70% clearly indicating that the epilithon is an active and highly viable community. Thus although the general features of epilithon are established there remains much work to be done to produce a better understanding of its physiology and microbial interactions.

Laboratory, microcosm and "in situ" experiments

Laboratory experiments are those done on agar under controlled conditions using a completely defined system. The use of mixed cultures, of unknown composition, i.e. suspensions of natural bacteria, in a laboratory based apparatus designed to mimic the environment, is an experiment done in a microcosm. So it is possible to design microcosms which vary in complexity and in the degree to which they represent the environment. True *in situ* experiments are those done with bacteria in an unenclosed manner in the environment, such that the natural flora and fauna can participate in the experiment. Thus in all experiments done *in situ* the environmental conditions are uncontrolled.

Exogenous isolation

Exogenous isolation (Fry and Day 1990) is a novel approach to the isolation of plasmids which selects natural plasmids initially on the basis of their ability to transfer. Phenotypic selection of the recipient and plasmid allows the transfer to be detected. Therefore plasmids, present in the natural community, can be identified by their transfer into marked recipient bacteria. This will identify the most actively transferring plasmids for the recipient species used (Fry and Day 1990).

Restriction

Restriction provides a significant barrier to the successful transfer of genes. In the laboratory it can reduce the frequency of exchange by at least 10^4-fold. Restriction systems of different specificity's are present in different strains and species and some have more than one (Wilkins 1990).

Description of methodology and strains

Strains and plasmid

The mercury resistance phenotype was chosen because it proved a most practicable selectable phenotype (Fry and Day 1991). It was a simple and effective plasmid encoded phenotypic marker that occurred infrequently in natural populations. Plasmid pQM1 (254 kb; Tra$^+$, Hg^r, FMA^r, Merbromin^r, Phi(E79), UV^r) was isolated exogenously from river epilithon and has been described previously

(Bale et al. 1987). Well studied laboratory strains and recently isolated epilithic strains were used for comparison. Mutants resistant to rifampicin and nalidixic acid were used for intra and interstrain crosses. The psychrotroph (*Pseudomonas fluorescens*, SAM) was characterised by an ability to form colonies on R2A (Reasoner and Geldrich 1985) at 4°C within 7d and its inability to form colonies at 37°C (Morita 1982). The mesophile (*Pseudomonas putida*, MES) grew at 37°C, but did not form colonies at 4°C.

In situ stone mating experiments

In the mating procedure (Bale et al. 1988a) stones, with or without (scrubbed) intact epilithon and with donor and recipient cultures on membrane filters, were held in nylon mesh bags and placed in the river. After 24 hours the stones were retrieved and transported to the laboratory, where donors, recipients and transconjugants were enumerated. Control experiments were also carried out to ensure conjugation only occurred in the river (Bale et al. 1987, 1988a).

Artificial stream microcosm

A recirculating stream microcosm (Vogel and Labarbera 1978) was used (Fig. 1).

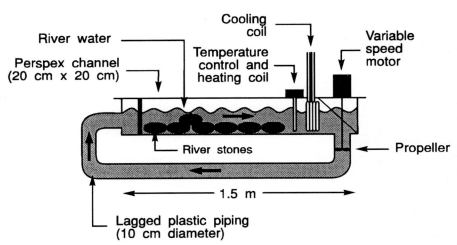

Fig. 1. Diagram of the recirculating stream microcosm.

Experiments and Discussion

Effect of temperature in laboratory, microcosm and "in situ" experiments

In these experiments two strains of *Pseudomonas aeruginosa* (donor: PAO2002 (pQM1); and recipient: PU21) were used. Membrane matings in the laboratory, in microcosms and *in situ* were done by several people over 5-6 years at various temperatures and the resulting transfer frequencies are shown in Figure 2. A comparison of the transfer frequencies obtained from the microcosm and *in situ* experiments shows they are generally similar. The same relationship with temperature is seen when these data are compared to the data from laboratory transfers, but the latter are significantly lower. The transfer profile for pQM1 shows that it transfers at a higher optimum frequency in the laboratory (greater than 10^{-1}) than it does either in a microcosm or *in situ*, on average at the same temperature (at 10^{-4} at 20-25°C). It is interesting that the data are so reproducible, despite the time and numbers of experiments; it implies a high consistency in the environment. The microcosm experiments were done in the recirculating stream system described earlier. The transfer frequencies were very similar to those obtained *in situ*, indicating the efficacy of this type of microcosm for predicting *in situ* results.

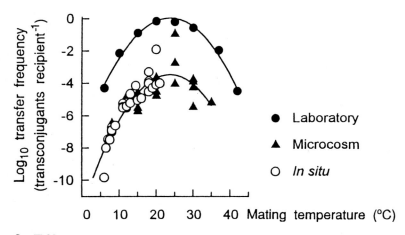

Fig. 2. Effect of temperature on the transfer of pQM1 from *Pseudomonas aeruginosa* PAO2002 to *P. aeruginosa* PU21.

The effect of physiological type and temperature on intra- and inter-strain pQM1 transfer

Two aspects were examined. Firstly cellular criteria were examined; those based on the different physiological types of the donor and recipient. Figure 3 shows the results from intra-strain mating experiments done at different temperatures (Diaper et al. 1992). In both cases the cells were grown at 20°C and then mated at various temperatures. The figure shows that although the mesophile has a broader transfer range in laboratory experiments than the psychrotroph, the psychrotroph transfers significantly better at 5°C.

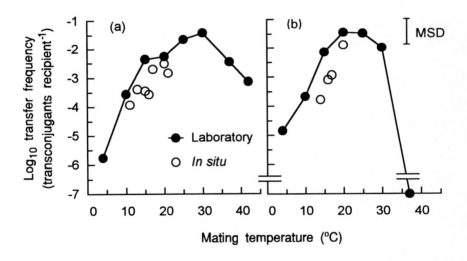

Fig. 3. Effect of temperature on the transfer of pQM1 within strains of (a) a mesophile (*Pseudomonas putida* MES) and (b) a psychrotroph (*Pseudomonas fluorescens* SAM). MSD, mean standard deviation.

Both are observations that might be expected from their respective growth physiology. In addition some experiments were done *in situ*. In these the frequencies of transfer obtained with the mesophile, while lower, were not significantly reduced. Those transfer frequencies obtained with the psychrotroph were only significantly reduced at temperatures below 20°C. Compare these transfer data with those obtained with laboratory strains of *P. aeruginosa* (Fig. 2), where there was a large significant difference, at all temperatures, between the *in situ* and laboratory frequencies (about 500-fold). In these crosses with natural epilithic isolates the differences were much smaller; 5-fold for the mesophile, which was not significant, and <20-fold for the psychrophile below 20°C.

Both growth and mating temperature effects transfer frequency (Table 1) for these two epilithic isolates. In self crosses, when grown at 20°C, the psychrotroph is 10-fold better at transferring pQM1 at 20°C than the mesophile, but both are good ($2-4 \times 10^{-2}$) at their respective optimum growth temperatures. Premating growth of the psychrotroph at 30°C reduces the efficiency of pQM1 transfer at 20-30°C by 10-100-fold, but has no significant effect on the mesophile.

Table 1. Effect of growth temperatures and host cell physiology[1] on the transfer of pQM1.

		Transfer frequency (transconjugants recipient^{-1}) after mating with the following donors grown at the stated temperature:			
	Mating	Psychrotroph		Mesophile	
Recipient	Tempera-ture (°C)	20°C	30°C	20°C	30°C
Psychrotroph	20	4×10^{-2}	5×10^{-4}	8×10^{-5}	4×10^{-5}
	30	1×10^{-2}	9×10^{-4}	1×10^{-3}	3×10^{-5}
	37	4×10^{-6}	2×10^{-4}	1×10^{-5}	4×10^{-6}
Mesophile	20	2×10^{-2}	6×10^{-4}	5×10^{-3}	5×10^{-3}
	30	2×10^{-2}	1×10^{-3}	4×10^{-2}	2×10^{-2}
	37	6×10^{-3}	8×10^{-4}	4×10^{-3}	3×10^{-3}

[1] Psychotroph = *Pseudomonas fluorescens* SAM;
Mesophile = *Pseudomonas putida* MES.

Again the transfer frequency of pQM1 at 37°C is severely reduced in the psychrotroph matings when grown prior to mating at 20°C. This is probably because the psychrotroph rapidly looses viability at 37°C. Growth of the psychrotroph at 30°C, prior to mating, allows habituation of the strain and permits it to transfer pQM1 equally well at each of the three temperatures tested.

Table 1 also shows the results of inter-strain pQM1 transfer experiments done between the mesophile and the psychrotroph. Growth prior to mating at 20°C is optimal for the psychrotroph, when it is the recipient, for matings done at 30°C. All other temperature combinations effectively reduce its efficiency by 10-250-fold. Intra-strain crosses with the mesophile as recipient show that growth of the psychrotroph donor at 20°C gives higher transfer frequencies than if it is grown at 30°C. Despite this the mating temperature has little effect on the ability of the psychrotroph to effectively transfer pQM1.

Figure 4 compares the results of crosses between the psychrotroph and the mesophile in laboratory and *in situ* experiments, all with both growth and mating at 20°C. These comparative experiments

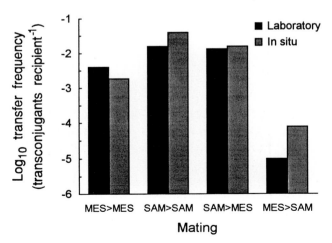

Fig. 4. Transfer of pQM1 between a mesophile (*Pseudomonas putida* MES) and a psychrotroph (*Pseudomonas fluorescens* SAM).

mostly show little difference in pQM1 transfer frequencies between these epilithic strains. Only in one direction is the transfer of pQM1 affected, and that is from the mesophile to the psychrotroph and it is reduced both in the laboratory and in *in situ* experiments. These results further confirm the similarity between laboratory and *in situ* results for these epilithic isolates.

The effects of restriction on pQM1 transfer
Some different experiments were done to establish the influence that restriction may have on pQM1 transfer. In the first, various epilithic strains of *P. fluorescens* and *P. putida* were used to study the effects of host species on pQM1 transfer. Figure 5 is a summary of these inter- and intra-strain data and shows that transfer within and into *P. putida* can be high, but that it does not extend to another closely related species *P. fluorescens*. When the median of these data are examined it is clear that there is a difference between the two fluorescent pseudomonad species.

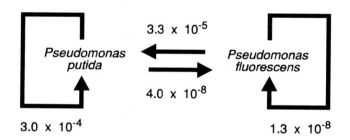

Fig. 5. Median laboratory transfer frequencies for pQM1 between 2 epilithic strains of the mesophile *Pseudomonas putida* and 5 of the psychrotroph *Pseudomonas fluorescens*.

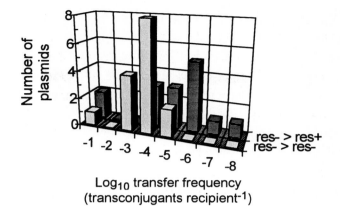

Fig. 6. Effect of restriction on transfer of pQM1 between fluorescent pseudomonads in the laboratory.

Figure 6 illustrates the effect of restriction on pQM1 transfer frequencies between restriction-proficient (res$^+$) and deficient strains (res$^-$) of fluorescent pseudomonads. It shows clearly that transfer into restriction-deficient strains occurs overall at a higher rate, but that transfer can be high in some crosses despite an effective restriction system. This suggests that additional factors (such as anti-restriction systems) are operating in determining the success of gene exchange.

Table 2. *In situ* matings between naturally occurring restriction-proficient and deficient *Pseudomonas* strains.

Strains	Lab (temp)	"In situ"[1] (temp)
Pseudomonas putida		
MES$^-$ X MES$^-$ (n=2)	2.6×10^{-2} (15°C)	1.0×10^{-3} (13°C)
MES$^-$ X DB10$^+$ (n=2)	5.3×10^{-5} (15°C)	1.6×10^{-7} (12°C)
Pseudomonas aeruginosa		
PAO 2002$^-$ x PU21$^+$ (n=2)	2.8×10^{-1} (15°C)	5.0×10^{-4} (15°C)

[1] Mean of duplicate experiments and transconjugants counted on stone and filter surface.

Other types of experiment were done *in situ*. In the first epilithic and laboratory strains were used and the donor strains were all restriction-deficient (Table 2). The frequency of pQM1 transfer from the epilithic donor strain (MES) was slightly reduced when laboratory and *in situ* experiments were compared. Transfer into a restriction deficient *P. putida* strain (from MES to DB10) showed a reduction in frequency, like that obtained from a similar *P. aeruginosa* cross. Thus restriction does have an effect on transfer *in situ*. The other *in situ* experiments involved the exogenous isolation of natural plasmids into four strains to determine the effects of restriction on the success of natural plasmid isolation. The restriction (res) status of these strains has been determined; three are epilithic isolates and the fourth (UWC1) is a laboratory strain. The restriction deficient (res$^-$) strains were clearly more effective, (median 1.6×10^{-2}) than the restriction-proficient (res$^+$) strains (median 3.5×10^{-5}) by a 1000-fold, at isolating natural plasmids from crude epilithon samples.

Conclusions

Overall the natural epilithic plasmid, pQM1, transfers at high frequencies both *in situ* and in laboratory matings using laboratory and natural fluorescent pseudomonads. Interestingly, the transfer of pQM1 between the natural epilithic isolates of *P. putida* and *P. fluorescens* are less effected by *in situ* incubation than are the laboratory strains. This suggests that the use of laboratory strains to predict transfer rates *in situ* will significantly underestimate transfer events.

The physiology of the strains involved in the transfer process are significant to the success of transfer of pQM1 in particular instances. The mesophile has an optimum transfer temperature around 25-30°C compared to that of the psychrotroph of about 22°C. The intra- and inter-species transfer experiments with the mesophile and psychrotroph show that both donor and recipient cell physiology and temperature can, but not always predictably, influence transfer success.

Restriction can clearly be a significant factor limiting transfer *in situ,* as it is in laboratory expriments, but its effects, like those of temperature and physiology, are not entirely predictable.

Acknowledgements

Most of our research described here has been done with support from the Natural Environmental Research Council, in the form of research grants and studentships, and from the European Economic Community under the *aegis* of the Biotechnology Action Programme. Thanks are also due to some of our research staff whose work is discussed here and some of whom have allowed us to present unpublished material. These are Dr B. Cousland and Dr D. Bradley.

References

Altherr MR, Kasweck KL (1982) *In situ* studies with membrane diffusion chambers of antibiotic resistance transfer in *Escherichia coli.* Appl Environ Microbiol 43:838-843

Aznar R, Amaro C, Alcaide E (1988) Characterisation of R-pL in the environmental isolates of *Salmonella*: Host range and stability. Curr Microbiol 17:173-177

Bale MJ, Fry JC, Day MJ (1987) Plasmid transfer between strains of *Pseudomonas aeruginosa* on membrane filters attached to river stones. J Gen Microbiol 133:3099-3107

Bale MJ, Fry JC, Day MJ (1988a) Transfer and occurrences of large mercury resistance plasmids in river epilithon. Appl Environ Microbiol 54:972-978

Bale MJ, Day MJ, Fry JC (1988b) Novel method for studying plasmid transfer in undisturbed river epilithon. Appl Environ Microbiol 54:2756-2758

Baya AM, Brayton PR, Brown VL, Grimes DJ, Russek-Cohen E, Colwell RR (1986) Co-incident plasmid and antimicrobial resistance in marine bacterial isolated from polluted and unpolluted Atlantic Ocean samples. Appl Environ Microbiol 51:1285-1292

Burton NF, Day MJ, Bull AT (1982) Distribution of bacterial plasmids in clean and polluted sites in a South Wales river. Appl Environ Microbiol 44:1026-1029

Cruz-Cruz NE, Toranzos GA, Ahearn DG, Chazen T (1988) *In situ* survival of plasmid-bearing and plasmid-less *Pseudomonas aeruginosa* in pristine tropical waters. Appl Environ Microbiol 54:2574-2577

Corliss TL, Cohen PS, Carbelli VJ (1981) R-plasmid transfer to and from *Escherichia coli* strains isolated from human faecal samples. Appl Environ Microbiol 41:959-966

Diaper J, Day MJ, Fry JC (1992) The effect of temperature and host cell physiology on the conjugal transfer of mercury resistance plasmid in the laboratory and river epilithon. J Gen Microbiol Submitted

Fry JC, Day MJ (1990) Plasmid transfer and the release of genetically engineered bacteria in nature: a discussion and summary. In; Fry JC, Day MJ (eds) Bacterial genetics in natural environments. Chapman and Hall, London, p. 243-250

Fry JC, Zia T (1982) Viability of heterotrophic bacteria in freshwater. J Gen Microbiol 128:2841-2850

Gauthier MJ, Cauvin F, Breittmayer JP (1985) Influence of salts and temperature on the transfer of mercury resistance from a marine pseudomonad to *Escherichia coli*. Appl Environ Microbiol 50:38-40

Gealt MA, Chai DM, Alpert KB, Boyer JC (1985) Transfer of plasmids pBR322 and pBR325 in wastewater from laboratory strains of *Escherichia coli* to bacteria indigenous to the wastewater disposal system. Appl Environ Microbiol 49:836-841

Genthner FJ, Chatterjee P, Barkay T, Bourquin AW (1988) Capacity of aquatic bacteria to act as recipients to plasmid DNA. Appl Environ Microbiol 115:117

Gowland P, Slater JH (1984) Transfer and stability of drug resistance plasmids in *Escherichia coli*. Microb Ecol 10:1-13

Hada HS, Sizemore S (1981) Incidence of plasmids in marine *Vibrio* spp. isolated from an oilfield in the north western Gulf of Mexico. Appl Environ Microbiol 41:199-202

Lock MA (1981) River epilithon: a light and organic energy transducer. In: Lock MA, Williams DD (eds) Perspectives in running water ecology. Plenum Press, New York

Lock MA, Wallace RP, Costerton JW, Ventullo RM, Charlton SE (1984) River epilithon: towards a structural and functional model. Oikos 42:10-22

Kelly WJ, Reanney DC (1984) Mercury resistance among soil bacteria: ecology and transferability of genes encoding resistance. Soil Biol Biochem 16:1-8

King GJ (1989) Plasmid analysis and variation in *Pseudomonas syringae*. J Appl Bacteriol 67:489-496

Kolenc R, Inniss WE, Glick BR, Robinson CW, Mayfield CL (1988) Transfer and expression of mesophilic plasmid mediated degradative capacity in psychrotrophic bacterium. Appl Environ Microbiol 54:638-641

Korbori H, Sullivan CW, Shizuya H (1984) Bacterial plasmids in Antartic natural assemblages. Appl Environ Microbiol 48:515-518

Mach PA, Grimes DJ (1982) R-plasmid transfer in a wastewater treatment plant. Appl Environ Microbiol 44:1395-1403

Morita RY (1982) Starvation-survival of heterotrophs in the marine environment. Adv Microb Ecol 6:171-198

O'Morchoe SB, Ogunseitan O, Sayler GS, Miller RV (1988) Conjugal transfer of R68.45 and FP5 between *Pseudomonas* strains in a freshwater environment. Appl Environ Microbiol 54:1923-1929

Reanney DC, Roberts WP, Kelly MJ (1982) Genetic interactions among microbial communities. In: Bull AT, Slater JH (eds) Microbial interactions and communities Vol 1. Academic Press, London, p. 287-322

Reasoner DJ, Geldreich EE (1985) A new medium for the enumeration and subculture of bacteria from potable water. Appl Environ Microbiol 49:1-7

Rochelle PA, Fry JC, Day MJ (1989) Factors effecting conjugal transfer of plasmids encoding mercury resistance from pure cultures and mixed natural suspensions of epilithic bacteria. J Gen Microbiol 135:409-424

Shaw DR, Carbelli VJ (1980) R-plasmid transfer frequencies from environmental isolates of *Escherichia coli* to laboratory and faecal strains. Appl Environ Microbiol 40:756-764

Singleton P, Anson AE (1981) Conjugal transfer of R plasmid R1 *drd* 19 in *E. coli* below 22°C. Appl Environ Microbiol 42:789-791

Smith W, Parsell Z, Green P (1978) Thermosensitive antibiotic resistance plasmids in Enterobacteria. J Gen Microbiol 109:37-47

Trevors JT, Oddie KM (1986) R-plasmid transfer in soil and water. Can J Microbiol 33:191-198

Van Es FB, Meyer-Reil LA (1982) Biomass and metabolic activity of heterotrophic marine bacteria. Adv Microbial Ecology 6:111-170

Vogel S, Labarbara M (1978) Simple flow tanks for research and teaching. Bioscience 28:638-643

Walmsley RH (1976) Temperature dependence of mating pair formation in *Escherichia coli*. J Bacteriol 126:222-224

Wilkins BM (1990) Factors influencing the dissemination of DNA by bacterial conjugation. In: Fry JC, Day MJ (eds) Bacterial genetics in natural environments. Chapman and Hall, London, p. 22-30

Wickham GS, Atlas RM (1988) Plasmid frequency fluctuations in bacterial populations from chemically stressed soil communities. Appl Environ Microbiol 54:2192-2196

Wortman AT, Colwell RR (1988) Frequency and characteristics of plasmids in bacteria isolated from deep sea Amphipods. Appl Environ Microbiol 54:1284-1288

Virus-Mediated Gene Transfer in Freshwater Environments

R.V. Miller, S. Ripp, J. Replicon[1], O.A. Ogunseitan[2] and T.A. Kokjohn[3]
Department of Microbiology and Molecular Genetics
Oklahoma State University
Stillwater, Oklahoma 74078
USA

Three major systems of genetic transfer are recognized in bacteria: conjugation, transduction, and transformation. These systems have been used routinely in the laboratory as tools of genetic analysis for many years, but their importance in investigating microbial genetic diversity and evolution in natural habitats has only recently begun to be investigated (Levy and Miller 1989). To date, most environmental studies have focussed on the potential for conjugation to transfer extra-chromosomal elements among microbes of the same or different species (Sayre and Miller 1991). Transduction (virus-mediated, horizontal gene transfer) has often been discounted as a potentially important process for the redistribution of genetic information (both chromosomal and extra-chromosomal) in bacterial populations because it is reductive (i.e., the donor is killed in the process of donating its genetic material to the recipient). However, recent reports have documented that transduction can be a fertile gene exchange system in natural ecosystems (Kokjohn 1989,

---

[1] Program in Molecular Biology, Loyola University of Chicago, Chicago, IL, USA
[2] Program in Social Ecology, University of California, Irvine, CA, USA
[3] Environmental Research Division, Argonne National Laboratory, Argonne, IL, USA

Kokjohn and Miller 1992). We are using *Pseudomonas aeruginosa* as a model organism to study virus-mediated gene transfer in freshwater microbial populations (Morrison et al. 1978; Saye et al. 1987; 1990; Saye and Miller 1989; Miller et al. 1990; Ripp et al. 1992). Our studies have revealed a significant potential for transduction of both plasmid and chromosomal DNA in these environments.

Initially, we used the well characterized, generalized transducing phage F116 (strain DS1) (Miller et al. 1974; 1976) to explore three models for the source of transducing particles in freshwater environments: (a) cell-free lysates of bacteriophages grown on an appropriate DNA donor, (b) environmentally induced bacteriophages from a lysogenic DNA donor bacterium, and (c) environmentally induced bacteriophages from a lysogenic recipient bacterium. The transfer of plasmid and chromosomal DNA was documented in each of these systems *in situ* in a freshwater lake (Saye et al. 1987; 1990). Experiments were carried out in environmental containment chambers (Saye and O'Morchoe 1992) into which genetically marked strains of *P. aeruginosa* were introduced. The chambers were filled with unaltered or sterilized lake water from the field site, and they were incubated near the surface of the lake for up to three weeks. The highest numbers of transductants were routinely recovered from systems in which the recipient bacterium was a lysogen, probably because of the immunity to superinfection imparted by the resident prophage. Transduction was observed in both the absence and presence of the natural microbial community (Saye et al. 1987). Reciprocal chromosomal transduction was observed in chambers inoculated with two lysogens (Saye et al. 1990). Apparently, both primary infection of a non-lysogen and spontaneous prophage induction from a lysogen can generate sufficient numbers of transducing particles to allow gene exchange to be observed.

These investigations predicted that the most likely reservoir of environmental transducing bacteriophages is the lysogenized members of the natural microbial population. This prediction has led us to formulate a model for the dispersal of genetic material from an introduced organism to related members of the autochthonous microbial community (Fig. 1). The model is equally applicable to the transport of chromosomal and extra-chromosomal DNA elements, but it is illustrated here for the transduction of a drug resistance plasmid introduced into the gene pool from a genetically engineered microorganism (GEM) or a natural organism transported into the habitat by natural or artificial means (Saye and Miller 1989).

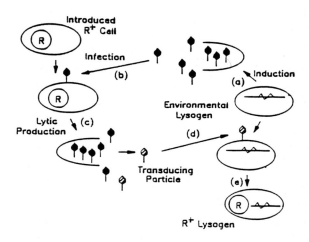

Fig. 1. Model of transduction of plasmid DNA among bacteria in an aquatic ecosystem. This model requires a unique sequence of events: (a) Phage virions must be produced through spontaneous or stress-stimulated activation of prophages from environmental lysogens. These viral particles must (b) infect, propagate, and (c) lyse the introduced DNA donor. (d) Transducing particles produced during this lytic infection must absorb and transfer DNA to the remaining lysogens, where the foreign DNA element is established either as a plasmid (e) or, if it is chromosomal in origin, is recombined into the host chromosome. Hence, environmental lysogens are both efficient sources of transducing phages and as viable recipients for transduced DNA (Reprinted with permission from Levy and Miller 1989).

The finding that transduction can occur in natural environments is significant because this mechanism of gene transfer has been virtually ignored in both the design and preliminary testing of GEMs for environmental release. However, the ultimate question that must be addressed in determining the effects of environmental transduction is whether phage-mediated horizontal gene transfer can alter the probability of retention or the equilibrium frequency of an introduced genetic sequence in a natural bacterial gene pool.

The potential of our transduction model to affect genetic diversity in bacterial populations was demonstrated in continuous-culture studies designed to determine the evolution of mixed genetic populations of the type described in the model (Replicon and Miller 1990). We found that at cell densities and generation times (hydraulic residence times) similar to those expected in aquatic environments, transduction stabilized and even increased the frequency of genotypes that were lost through negative selection in continuous cultures where gene transfer was disallowed. In experiments conducted at similar equilibrium cell densities but different hydraulic residence times, we found that the rate of transduction was directly proportional to the phage-to-bacterium ratio (PBR) developed in the microbial consortium. The PBR was, in turn, proportional to the generation time and, hence, to the extent of starvation of the host organism. Lower equilibrium concentrations of phage virions were produced per bacterial cell when the turnover time of the chemostat was slow than when it was fast. This is an exciting and significant result because transduction has often been discounted as an important evolutionary because it is perceived to be reductive.

Thus, our studies suggest that transduction can occur in natural habitats and can significantly influence the genetic makeup and diversity of natural bacterial populations. To determine the significance of this method of gene transfer in natural ecosystems, one must understand the dynamics of phage-host interactions and identify the reservoirs of environmentally observed bacteriophage particles. Such studies are also necessary to determine the potential of phages to control and limit natural and introduced populations of their bacterial hosts in the environment.

To begin an analysis of phage-host interactions in the aquatic environment, we monitored the occurrence of bacteriophages, potential hosts, and lysogens at one of our freshwater field sites over a nine-month period (Fig. 2). This study indicated that phages

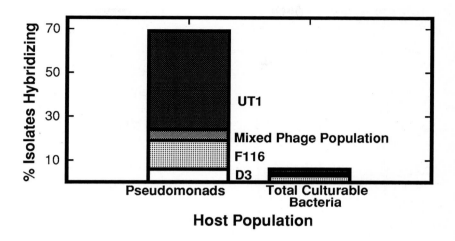

Fig. 2. Abundance of bacteria and *Pseudomonas*-specific bacteriophages in a freshwater habitat during a nine-month period. Total recoverable colony-forming units (CFU) (Δ) were plated on 0.1 X YEPG (Kokjohn et al. 1991). CFUs of *Pseudomonas* sp. (◊) were determined by plating on *Pseudomonas* Isolation Agar. Plaque-forming units (PFU) of phages specific for *P. aeruginosa* (•) were identified by their ability to form plaques on laboratory strains.

capable of producing plaques on an environmentally prominent bacterial host are present in significant titers in freshwater habitats (Replicon and Miller 1990). When lysogeny was evaluated, between 1-7% of the *P. aeruginosa* isolates tested positive by the criterion of release of plaque-forming units (PFU) infectious on laboratory strains of *P. aeruginosa*. In another field study (Ogunseitan et al. 1990; 1992), we used colony hybridization to assess the frequency of lysogeny in natural *P. aeruginosa* isolates. The use of various DNA probes specific for *P. aeruginosa* phages of environmental origin revealed that a large fraction (70%) of *Pseudomonas* isolates contained phage-specific DNA (Fig. 3). These data indicate that the frequencies of *Pseudomonas* sp., *Pseudomonas*-specific phages, and lysogens are high in natural freshwater ecosystems. They also suggest that significant environmental reservoirs of bacteriophages have gone uncharacterized and must be explored in greater detail before a realistic picture of the importance of phages to microbial ecology can be obtained.

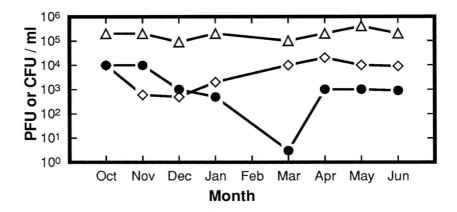

Fig. 3. Relative abundance of sequences related to *Pseudomonas* phage DNA in bacterial colonies recovered from a freshwater habitat. Recoverable colony-forming units from a freshwater lake were plated on 0.1 X YEPG for total bacteria or on *Pseudomonas* Isolation Agar for pseudomonads. Colony hybridization was carried out by using phage-specific probes. (Data summarized from Ogunseitan et al. 1990, 1992.)

The potential for transduction to increase the genetic diversity of these microbial communities depends on the ability of bacteriophages and host bacteria to interact in natural habitats. Very little information is currently available on this subject (Miller and Sayler 1992). We have initiated studies to investigate the natural reservoirs of bacteriophages and the dynamics of phage-host interactions in the aquatic environment. Our preliminary observations suggest that this evaluation will provide important and sometimes surprising insights into microbial population ecology. The data we have collected to date indicate that bacteriophages can productively interact with their hosts at concentrations and under conditions found in nature.

We have isolated several *P. aeruginosa* bacteriophages from our freshwater field sites and studied the dynamics of the interaction of one, UT1, with its environmental hosts in some detail (Ogunseitan et al. 1990; 1992; Ripp et al. 1992). When isolates of *P. aeruginosa* from the field site were examined, both UT1-sensitive and UT1-resistant strains were identified.

The attachment and replication of the *P. aeruginosa* bacteriophages F116 and UT1 were investigated under conditions simulating aquatic environments (Kokjohn et al. 1991). These phage-host interactions were not impaired at host cell densities equal to or lower than those frequently found in aquatic environments when the host cells were physiologically competent to allow phage growth. When hosts cells were starved in river water, attachment was not impaired, but replication of the bacteriophages was significantly altered. In primary infection, the latency period was lengthened and the burst size reduced in comparison to fed cells (Table 1).

We studied the interaction of UT1 and newly isolated *P. aeruginosa* hosts indigenous to a freshwater lake during a 45-day incubation in lake water microcosms (Ogunseitan et al. 1990). Temporal changes in host density, PBR, and the appearance of apparent prophage carriers within the host population were monitored. Decay rates of the phage virion ranged from 0.054 h^{-1} in natural water to 0.027 h^{-1} in filter-sterilized lake water. Bacterial densities stabilized at significantly lower densities when UT1 was present (10^4 CFU/ml) than when it was absent (10^7 CFU/ml), but about 45% of the sensitive bacteria incubated with UT1 hybridized to probes of UT1-specific DNA sequences within 72 h of incubation in sterile lake water and within 12 h in microcosms containing natural lake water. Plaques formed by this phage upon initial isolation were turbid, but only clear plaques were observed when the phages were plated on laboratory-grown hosts. These data suggest that although UT1 is apparently virulent on well fed hosts, starvation conditions found in the aquatic habitat favor the establishment of a pseudo-lysogenic relationship between the bacteriophage and its host bacteria, like the one described by Romig and Brodetsky (1961) in soil bacilli.

Table 1: Effect of starvation on the latent period (LP) and burst size (BS) of two *Pseudomonas aeruginosa* bacteriophages.

Growth condition	Phage	LP (min)	BS (PFU)
Fed	F116	100-110	27 ± 4
	UT1	70-80	65 ± 13
Starved 48 h	F116	≥ 240	4 ± 2
	UT1	≥ 110	6 ± 13

Pseudomonas aeruginosa strain PAO303 (*argB21*) was used for these studies. Fed cells were prepared by growing in Luria Broth. Starved cells were prepared from Luria Broth cultures in early exponential phase by dilution (1:10) in autoclaved river water obtained from the Des Plaines River in Maywood IL and incubated at 20°C for the period of time indicated. (Data summarized from Kokjohn et al. 1991.)

   The ability of UT1 to mediate generalized transduction of both chromosomal and extra-chromosomal DNA was assessed (Ripp et al. 1992). Bacterial cells from 10 l of lake water were concentrated and used as a recipient pool for UT1-mediated transduction of the Mob⁻ Tra⁻ plasmid Rms149 (Saye et al. 1987), which confers carbenicillin (Cb) and streptomycin (Sm) resistance. Few Cb[r] and Sm[r] clones were observed among the lake water community, but the frequency of these colonies increased as much as 100 fold after transduction by a UT1 lysate grown on an Rms149-containing donor strain (Ripp et al. 1992). These data indicate that all of the components necessary for transduction are present in representative freshwater microbial communities.

   We are now turning our attention to developing methods to test *in situ* a model transduction system whose components are derived directly from the environment. These studies will allow us to evaluate of the effectiveness of transduction as a mechanism for altering the makeup of the gene pool available to natural populations of bacteria. They will provide a more accurate assessment of the real potential for genetic exchange to affect natural microbial populations than is currently possible with systems using laboratory-derived donor and recipient bacteria.

Transduction of genetic material has now been verified as occurring *in situ* in waste water treatment facilities (Osman and Gealt 1988), in soils (Zeph et al. 1988; Stotzky 1989; Zeph and Stotzky 1989), in animals (Jarolmen et al. 1965; Novick and Morse 1967; Baross et al. 1974; Novick et al. 1986), on plant surfaces (Kidambi et al. 1992), and in freshwater environments (Morrison et al. 1978; Saye et al. 1987; 1990; Amin and Day 1988; Saye and Miller 1989; Miller et al. 1990; Ripp et al. 1992). These findings have revealed that transduction can take place in natural environments. They raise important questions about the contributions of transduction to the diversity of microbial gene pools and the evolution of natural microbial communities and to the risk associated with gene movement from GEMs to the microbiota with which they may interact after release into natural environments. Acceptable answers to these questions will be obtained only with further insight into the characteristics of the interactions between bacteriophages and bacteria in natural ecosystems.

Acknowledgements

This work was supported in part by cooperative agreement CR818254 with the Environmental Research Laboratory, Gulf Breeze, Florida, USA, of the US Environmental Protection Agency and in part by the US Department of Energy, Office of Energy Research, Office of Health and Environmental Research, under contract W-31-109-Eng-38.

References

Amin MK, Day MJ (1988) Donor and recipient effects on transduction frequency *in situ*. REGEM1 Program, p. 2.

Baross JH, Liston J, Morita RY (1974) Some implication of genetic exchange among marine vibrios, including *Vibrio parahaemolyticus*, naturally occurring in the Pacific oyster. In Fujio T, Sakaguchi G, Sakazaki R, Takeda Y (eds.) International Symposium on *Vibrio parahaemolyticus*. Saikon Pub. Co. Ltd., Tokyo, pp. 129-137.

Jarolmen H, Bonke A, Crowell RL (1965) Transduction of *Staphylococcus aureus* to tetracyclin resistance *in vivo*. J Bacteriol 89:1286-1290.

Kidambi SP, Ripp S, Miller RV (1992) Evidence for phage-mediated gene transfer among *Pseudomonas aeruginosa* on the phylloplane. Submitted.

Kokjohn TA (1989) Transduction: mechanism and potential for gene transfer in the environment. In Levy SB, Miller RV (eds.), Gene Transfer in the Environment. McGraw-Hill, New York, pp. 73-98.

Kokjohn TA, Miller RV (1992) Gene transfer in the environment: transduction. In Fry JC, Day MJ (eds.), Release of Genetically Engineered and Other Microorganisms, Edward Arnold, London, in press.

Kokjohn TA, Sayler GS, Miller RV (1991) Attachment and replication of *Pseudomonas aeruginosa* bacteriophages under conditions simulating aquatic environments. J Gen Microbiol 137:661-666.

Levy SB, Miller RV (1989) Gene Transfer in the Environment. McGraw-Hill, New York.

Miller RV, Kokjohn TA, Sayler GS (1990). Environmental and molecular characterization of systems which affect genome alteration in *Pseudomonas*. In Silver S, Chakrabarty AM, Iglewski B, Kaplan S (eds.), *Pseudomonas:* Biotransformations, Pathogenesis, and Evolving Biotechnology. American Society for Microbiology, Washington, D. C., pp. 252-268.

Miller RV, Pemberton JM, Richards KE (1974) F116, D3, and G101: temperate bacteriophages of *Pseudomonas aeruginosa*. Virology 59:566-569.

Miller RV, Pemberton JM, Clark AJ (1976) Prophage F116: evidence for extrachromosomal location in *Pseudomonas aeruginosa* strain PAO. J Virol 22:844-847.

Miller RV, Sayler GS (1992) Bacteriophage-host interactions in aquatic systems. In Wellington EM, van Elsas JD (eds.), Genetic Interactions Among Microorganisms in the Natural Environment, Pergamon Press, Oxford, UK, pp. 176-193.

Morrison WD, Miller RV, Sayler GS (1978) Frequency of F116 mediated transduction of *Pseudomonas aeruginosa* in a freshwater environment. Appl Environ Microbiol 36:724-730.

Novick RP, Edelman I, Lofdahl S (1986) Small *Staphylococcus aureus* plasmids are transduced as linear multimers that are formed and resolved by replicative processes. J Mol Biol 192:209-220.

Novick RP, Morse SI (1967) *In vivo* transmission of drug resistance factors between strains of *Staphylococcus aureus*. J Expt Med 125:45-59.

Ogunseitan OA, Sayler GS, Miller RV (1990) Dynamic interaction of *Pseudomonas aeruginosa* and bacteriophages in lake water. Microb Ecol 19:171-185.

Ogunseitan OA, Sayler GS, Miller RV (1992) Application of DNA probes to the analysis of bacteriophage distribution patterns in the environment. Appl Environ Microbiol, in press.

Osman MA, Gealt MA (1988) Wastewater bacteriophages transduce genes from the chromosome and a recombinant plasmid Abst. Annl Meet Am Soc Microbiol p. 254.

Replicon J, Miller RV (1990) Modeling the potential for transduction to stabilize a foreign genotype within an established microbial community. Abst VIIIth Internat Cong Virol, p. 117.

Ripp S, Ogunseitan OA, Miller RV (1992) Transduction of a freshwater microbial community by a new *Pseudomonas aeruginosa* generalized transducing phage, UT1. Submitted

Romig WR, Brodetsky AM (1961) Isolation and preliminary characterization of bacteriophages of *Bacillus subtilis*. J Bacteriol 82:135-141.

Saye DJ, Miller RV (1989) Gene transfer in aquatic environments. In Levy SB, Miller RV (eds.), Gene Transfer in the Environment, McGraw-Hill, NewYork, p. 223-254.

Saye DJ, Ogunseitan OA, Sayler GS, Miller RV (1990) Transduction of linked chromosomal genes between *Pseudomonas aeruginosa* during incubation *in situ* in a freshwater habitat. Appl Environ Microbiol 56:140-145.

Saye DJ, Ogunseitan O, Sayler GS, Miller RV (1987) Potential for transduction of plasmids in a natural freshwater environment: effect of plasmid donor concentration and a natural microbial community on transduction in *Pseudomonas aeruginosa*. Appl Environ Microbiol 53:987-995.

Saye DJ, O'Morchoe SB (1992) Evaluating the potential for genetic exchange in natural freshwater environments. In Levin M, Seidler R, Rogul M (eds.), Microbial Ecology: Principles, Methods, and Application in Environmental Biotechnology. McGraw-Hill, NewYork, pp. 283-309.

Sayre PG, Miller RV (1991) Bacterial mobile genetic elements: importance in assessing the environmental fate of genetically engineered sequences. Plasmid 26:151-171.

Stotzky G (1989) Gene transfer among bacteria in soil. In Levy SB, Miller RV (eds.), Gene Transfer in the Environment, McGraw-Hill, NewYork, pp. 165-222.

Zeph LR, Stotzky G (1989) Use of a biotinylated DNA probe to detect bacteria transduced by bacteriophage P1 in soil. Appl Environ Microbiol 5:661-665.

Zeph LR, Onaga MA, Stotzky G (1988) Transduction of *Escherichia coli* by bacteriophage P1 in soil. Appl Environ Microbiol 54:1731-1737.

Intergeneric Natural Plasmid Transformation between *Escherichia coli* and a Marine *Vibrio* Species

J.H. Paul
Department of Marine Science
University of South Florida
140 Seventh Ave. S.
St. Peterburg, Florida 33701
USA

Introduction

Because of the use of genetically engineered organisms in the environment, these has been renewed interest in transformation as a mechanism for horizontal gene transfer to the indigenous flora (Lorenz et al. 1988 ; Stewart and Sinigalliano 1990 ; Paul et al. 1991b). The potential for intergeneric, donor-mediated plasmid transfer to occur has not been widely investigated. There have been several reports of transfer of plasmid DNA from *Escherichia coli* cells to *Bacillus subtilis* (Van Randen and Venema 1984; Kosovich and Prozorov 1991). In this paper we describe the donor mediated transfer of plasmids between *E. coli* and *Vibrio* JT-1, the first report of intergeneric transformation involving a marine bacterium.

Materials and Methods

The plasmid pQSR50, a broad host range derivative of R1162 containing Tn5, was used as transforming plasmid, and was maintained in the donor, *E. coli* RM1259. All *Vibrio* recipient strains were derivatives of high frequency of transformation (HfT) *Vibrio* strains (Frischer et al. 1990). JT-1 was a spontaneous nalidixic acid, rifampin-resistant mutant of WJT-IC. All *E. coli* and *Vibrio* strains were grown as previously described (Frischer et al. 1990) except JT-1, which was grown in the presence of 500 µg/ml naladixic acid, and 150 µg/ml rifampicin.

For transformation assays, equal volumes of washed, overnight cultures of donors and recipients were mixed in 15 ml sterile tubes (Liquid assays) or comixed and filtered through 0.2 µm Nuclepore filters. Cells were coincubated and plated on the appropriate selective media (see above). Transformants were plated on ASWJP containing kanamycin, streptomycin, naladixic acid and rifampin. For both filter or liquid assays, DNase controls were performed by addition of 200 Kunitz units of DNase 1 (Sigma Chemical Co.). Plasmid acquisition was verified by colony hybridization and restriction analysis of minipreps using a probe made from the neomycin phosphotransferase gene (*npt*II) of Tn5 of pQSR50 (Frischer et al. 1990). Transformation assays with purified plasmid multimers were performed as previously described (Frischer et al. 1990).

Results and Discussion

Table 1 shows the results of studies using *E. coli* RMI259 as a plasmid donor for transformation of *Vibrio* JT-1. Transfer frequencies per recipient were higher in liquid (2×10^{-6} transformants/recipient) than on filters (5×10^{-7}). No spontaneous mutation of either the donor cells to nalidixic and rifampin resistance, nor of the recipients to kanamycin and streptomycin resistance was observed. The transfer was DNase-sensitive, and DNase controls in all experiments yielded no transfer (data not shown for all experiments). The results of colony hybridization studies on putative JT-1 transformants indicated that all such colonies hybridized to the probe, while the native JT-1 did not (data not shown).

Heat killed donor cells yielded transformation frequencies (per recipient) nearly identical to those obtained with viable donor cells (Table 1). Transformation assays performed in the presence of nalidixic acid and rifampicin resulted in transfer frequencies nearly two orders of magnitude below that obtained in the absence of these inhibitors (Table 1). If donor and recipient cells were separated by a 0.2 μm filter, no transfer occurred, in both filter and liquid assays (Table 1). These observations indicate that cell contact is required for plasmid transfer.

We have described inter- and intrageneric transfer of nonconjugative plasmids by cell contact-mediated transformation in the marine *Vibrio* JT-1. This process resembled conjugation in that cell contact was required. Unlike conjugation, the process was DNase sensitive (and by definition, transformation) and involved nonconjugative plasmids.

Table 1. Natural plasmid transformation of *Vibrio* JT-1

Treatment	Form of trans- forming DNA	Filter (F) or Liquid (L)	Transformation Frequency (F ± SD)
RM1259XJT-1	Viable don.cells	L	$2\pm0.6\times10^{-6}$
RM1259XJT-1	Viable don.cells	F	$5.2\pm3.7\times10^{-7}$
JT-1 only	none	F, L	$<8.5\times10^{-10}$
RM1259 only	none	F, L	$<6.8\times10^{-9}$
RM1259xJT-1 + DNase	Viable don.cells	F,L	$<3.1\times10^{-10}$
RM1259XJT-1	Heat killed cells	L	$2.7\pm2.5\times10^{-6}$
RM1259xJT-1	Heat killed cells	F	$2.7\pm3.2\times10^{-7}$
RM1259xJT-1 in Nal/Rif media	Donor cells impaired	L	1.9×10^{-8}
Nal/Rif media	Heat killed cells	L	8.2×10^{-7}
RM1259xJT-1 separated by 0.2 μm filter	Viable don.cells	F,L	$<9.3\times10^{-10}$
RM1259xJT-1 separated by 0.2μm filter	Heat killed cells	L	$<9.6\times10^{-10}$
JT-1 only	4μg plasmid mult	L	$3.0\pm2.2\times10^{-7}$
JT-1 only	4μg plasmid mult	F	1.0×10^{-5}
JT-1	Plasmid extract from 1 ml culture	F	$<7\times10^{-10}$

Unlike transformation with purified plasmids, transfer using donor cells was more efficient (per recipient) in liquid than in filter assays. Most mechanisms of transfer proceed more efficiently on surfaces than in liquid (Stewart and Carlson 1986 ; Rochelle et al. 1988). The presence of the donor cells apparently somehow facilitated transfer in liquid in contrast to DNA free in solution.

The absence of transfer when spent media was used as a source of transforming DNA or when donors and recipients were separated by a 0.2 μm filter indicated that appreciable concentrations of transforming plasmids were not released from the surface of the RMI259 donor cells. This perhaps explains the requirement for cell contact when donor cells were used as a source of transforming DNA.

The transfer that occurred when heat-killed donors were employed as a source of transforming DNA indicated that donor viability was not required for transfer. The inhibition of transfer in the presence of nalidixic acid and rifampicin, however, implies a metabolic function in the donor process. We hypothesize that active export of plasmids from the inside of the donor cells to the cell surface is required for donor mediated trasnfer. We also explain the transfer that occurred with heat killed cells as being the result of heat-induced conformational/structural changes in the donor, such that intracellular plasmids were released to the cell surface.

Our work demonstrates the potential for contact-mediated cell transfer to occur in the environment between widely differing genera. This may be a mechanism to disseminate nonconjugative, broad host range plasmids in aquatic environments. These results are relevant to situations where viable or nonviable E. coli or other enterics are released into the environment, as in treated and untreated sewerage outfalls. Our results may be relevant to understanding the fate of plasmids found in genetically modified organisms in the environment. The lack of viable donor cells (as determined by plating) does not preclude the portential for horizontal transfer by transformation.

This work was supported by NSF grant OCE 8817172 to JHP, a Gulf Coast Charitable Trust Foundation award and a Clearwater Power Squadron award to MEF.

References

Frischer MF, Thurmond JM, Paul JH (1990) Natural plasmid transformation in a high frequency of transformation *Vibrio* strain. Appl Environ Microbiol 56:3439-3444

Kosovich PV, Prozorov AA (1991) Intraspecies and interspecies transfer of plasmid and chromosomal DNA during natural transformation in *Bacillus subtilis* and *Escherichia coli*. Microbiol (USSR) 59:1046-1049.

Lorenz MG, Aardema BW, Wackernagel W (1988) Highly efficient genetic transformation of *Bacillus subtilis* attached to sand grains. J Gen Microbiol 134:107-112

Maniatis T, Fritsch EF, Sambrook J (1982) *Molecular cloning : a laboratory manual*. Cold Spring Harbor Laboratory, Cold Spring Harbor, NY

Paul JH, Frischer ME, Thurmond JM (1991) Gene transfer in marine water column and sediment microcosms by natural plasmid transformation. Appl Environ Microbiol 57:1509-1515

Rochelle PA, Day MJ, Fry JC (1988) Occurrence, transfer and mobilization in eptilithic strains of *Acinetobacter* of mercury-resistance plasmids capable of transformation. J Gen Microbiol 134:2933-2941

Stewart GJ, Carlson CA (1986) The biology of natural transformation. Ann Rev Microbiol 40:211-235

Stewart GJ, Sinigalliano CD (1990) Detection of horizontal gene transfer by natural transformation in native and introduced species of bacteria in marine and synthetic environments. Appl Environ Microbiol 56:1818-1824

Van Randen J, Venema G (1984) Direct plasmid transfer from replication of *E. coli* colonies to competent *B. subtilis* cells. Mol Gen Genet 195:57-61

Natural Transformation on Agar and in River Epilithon

H.G. Williams, M.J. Day and J.C. Fry
School of Pure and Applied Biology
University of Wales College of Cardiff
PO Box 915
Cardiff CF1 3TL
UK

Introduction

Natural transformation is a process in which competent cells take up and express exogenous DNA. It is one potential mechanism by which genetic sequences may be transferred through a natural population. There are several review articles discussing the importance of understanding genetic exchange in the environment (Levy and Miller 1989; Coughter and Stewart 1989; Fry and Day 1990). Conjugation has been shown to occur *in situ* in river epilithon (Bale et al. 1987). Transduction has been demonstrated in diffusion chambers in lake water (Saye et al. 1990) and transformation, in non sterile sediment microcosms (Stewart and Sinigalliano 1990). However, transduction and transformation have not yet been demonstrated, unenclosed *in situ* in natural environments. The aims of this study were to determine whether genetic information could be transferred by transformation in nature and to develop a simple procedure to determine the frequency of transfer *in situ*. We used *Acinetobacter calcoaceticus* because it is naturally competent and the genus is common in aquatic habitats (Baumann 1968).

Materials and Methods

Media and bacterial strains

Liquid cultures were grown in 4 ml of Luria broth (LB) at 20°C. Strains were subcultured on standard plate count agar (PCA, Oxoid, CM463). Minimal media were based on sodium succinate (10 g/l) in B22 salts solution (Bale et al. 1987) solidified with agar (10 g/l). Selective media were supplemented with antibiotics, mercury (Hg, 27 μg/ml) or amino acids (25 μg/ml). Antibiotics were normallly added at the following concentrations: rifampicin (Rif), 100 μg/ml; spectinomycin (Sp), 100 μg/ml; kanamycin (KM), 50 μg/ml; naladixic acid (Nal), 100 μg/ml.

Spontaneous mutants of *A. calcoaceticus* BD413 were isolated by plating LB cultures (100 μl) of the wild type on to selective media. A non-conjugative, mercury resistance plasmid, pQM17 (Rochelle et al. 1988) was introduced to some of the strains used (Table 1) by transformation.

Table 1. *Acinetobacter calcoaceticus* strains used and their phenotypes.

Strain	Phenotype [1]
BD413	Wild type
BD413(pQM17)	Hg^r
BD413R	Rif^r
HGW1501	Rif^r, Sp^r, His$^-$
HGW1521(pQM17)	Hg^r, Rif^r, Sp^r, His$^-$
HGW98(pQM17)	Hg^r, Met$^-$
HGW1017	Km^r
HGW1730	Nal^r

[1] Hg^r, mercury resistant; Rif^r, rifampicin resistant;
Sp^r, spectinomycin resistant; His$^-$, histidine auxotroph;
Met$^-$, methionine auxotroph; Km^r, kanamycin resistant;
Nal^r, naladixic acid resistant.

Transformation on agar in the laboratory
The source of transforming DNA was prepared either as an untreated LB culture or a crude lysate (Juni 1972). First the source of DNA (20 µl) then the recipient (20 µl, LB culture) were put on to the same spot on a PCA plate. The mixture was incubated at 20°C for 24 h, resuspended in B22 salts containing DNase I (50 µg/ml, Sigma, Poole, Dorset), and enumerated on selective media by drop counts. The transformation frequency was calculated as the number of transformants per recipient.

Transformant colonies were picked off and subcultured on to PCA. They were then screened for secondary characteristics by their ability to grow on appropriate selective media.

Testing selective media and in situ transformation
A suspension of epilithon was harvested from stones in the River Taff by scubbing (Bale et al. 1987). The epilithic suspension was serially diluted and spread in 100 µl aliquots on to various media. Samples were incubated at 20°C and colonies were counted after 96 h.

The source of DNA (1 ml) and recipient culture (1 ml) were deposited on to separate nitrocellulose filters (Whatman, 0.45 µm, 25 mm diameter), then transported on ice to the River Taff and placed together so that the recipient and source of DNA were in contact. The filters were placed on the surface of a sterile, scrubbed stone (Bale et al. 1987), and secured with a larger filter (Whatman No. 1) with elastic bands. The stones were then placed in a nylon mesh bag (mesh size 2 x 2 cm) which was attached by nylon monofilament to a metal stake and submerged, approximately mid stream on the bed of the river. After 24 h incubation the stones were removed and the filters were resuspended in 3 ml of B22 salts solution containing DNase I (50 µg/ml^{-1}). Samples were returned to the laboratory on ice and transfer frequencies were calculated as for laboratory experiments.

Results and Discussion

Transformation on agar in the laboratory

Recipient strains of *A. calcoaceticus* were shown to acquire altered genotypes following the addition of a source of transforming DNA. The source of DNA could be added as untreated broth cultures or crude lysates. To select an optimum transformation system for work *in situ,* using lysates and whole cells as DNA source, we compared transformation frequencies on PCA at 20°C using several genetic traits and appropriate mutants as recipients. The characteristics tested included auxotrophic markers (histidine, methionine, tryptophan, cysteine and leucine), antibiotic resistances (rifampicin, kanamycin, spectinomycin, erythromycin, naladixic acid, streptomycin) and mercury resistance.

Transformation frequencies were observed between 1.8×10^{-2} transformants per recipient (transformation to kanamycin resistance by a lysate of HGW1017) to $< 1 \times 10^{-9}$ transformants per recipient (transformation to naladixic acid resistance by a lysate of HGW1730). Transformation of a histidine auxotroph (HGW1501) to prototrophy by lysates of the wild type (BD413) was shown to be the most reproducible and stable event tested, with a high transfer frequency (7.59×10^{-4} transformants per recipient). Higher transfer frequencies were observed when lysates were used to transform the recipient than with untreated cultures of donor cells.

When lysates were used as the source of transforming DNA the gene being transferred and direction of transfer was unambiguous. To be sure of exactly which transfers were occurring in cell to cell matings it is important to ascertain the direction of transfer. So in experiments in which two strains were mixed, secondary, non-selected, characteristics were used to determine directionality by screening transformants. This approach established that transformation in each direction occurred at different frequencies. For example it was previously suggested that BD413R (a rifampicin resistant strain) was transformed to mercury resistance by an untreated culture of BD413(pQM17) at a frequency of 6×10^{-4} (Rochelle et al. 1988). Transformants were selected on PCA + Hg + Rif. However, our results with similar strains containing secondary auxotrophic markers (HGW98(pQM17) and HGW1501) showed that approximately 95 % of the transformants resulted from transformation of rifampicin resistance to the strain containing pQM17. Plasmid pQM17 was shown to transform at a frequency of only 5.38 x 10^{-7}.

Media for detecting transformants in epilithon
In experiments involving natural populations, the minimum number of transformants which could be detected was dependant on the dilution at which single colonies would still be countable and not overgrown by indigenous organisms. Table 2 lists the number of indigenous organisms per cm^2 capable of growth on various media used in our transformation studies to count transformants, donors or recipients.

Table 2. Enumeration of epilithic bacteria on selective media.

Media	Numbers (cfu per cm^2 stone surface)
PCA	1 x 10^7
PCA + Hg	1 x 10^5
PCA + Rif	2 x 10^3
PCA + Km	4 x 10^4
PCA + Sm (500 µg/ml)	3 x 10^3
PCA + Sp (75 µg/ml) + Rif (75 µg/ml) + Sm (2 µg/ml)	4 x 10^3
PCA + Hg + Rif	6 x 10^2
Succinate minimal medium	1 x 10^5
Succinate minimal medium + Hg (14 µg/ml)	10-30
Succinate minimal medium Rif (75 µg/ml)	8 x 10^2
Succinate minimal medium + Hg (14 µg/ml) + Rif (75 µg/ml)	0-5

The number of transformants per cm^2 has to be higher than the number of indigenous organisms forming colonies on the selective media to be countable. From laboratory experiments described above we expected to have less than 1×10^3 transformants per cm^2 in *in situ* experiments. The above results suggest that of the media tested only PCA + Hg + Rif, succinate minimal medium + Hg and succinate minimal medium + Hg + Rif would be useful for selecting transformants for *in situ* experiments. All these media were also very efficient at recovering transformants and gave 100 % recovery of pure cultures compared to PCA. Succinate minimal medium + Hg (14 µg/ml) + Rif (75 µg/ml) (S22) allowed the detection of transformants able to grow on this media at densities as low as 1 cell/cm^2 in natural epilithon. To ensure correct transformants counts in experiments involving mixtures of auxotrophs it was necessary to add EDTA (300 µg/ml) and succinate-B22 culture filtrable (0.5 % v/v) to S22. The EDTA prevented transformation on the selective media and the culture filtrate was required to ensure 100 % recovery of transformants (unpublished results). This medium was called SE22.

Detecting transformation in situ
An ideal system for examining transformation *in situ* must have both a high transfer frequency and an efficient medium to detect transformants on which indigenous bacteria grow poorly. The best system was transformation of a histidine auxotroph, HGW1521(pQM17) (Hg^r, Rif^r and Sp^r) to prototrophy by either a lysate or an untreated culture of BD413. Recipients were selected on PCA + Rif and transformants were selected on S22 when lysates were used or SE22 when untreated donor cultures were used. Laboratory transfer frequencies were 7.08×10^{-4} and 9.55×10^{-5} respectively for lysates and whole cells.

Transformation in the River Taff
HGW1521(pQM17) was transformed to prototrophy by either a crude lysate or untreated cultures of BD413 whilst immobilized on filters in the River Taff. Table 3 shows data from a typical experiment.

Table 3. Transformation of HGW1521(pQM17) to prototrophy

Source of trans-forming DNA	Recipient numbers (cfu/cm^2)	Transformant numbers (cfu/cm^2)	*In situ* transfer frequency per recipient
Lysate of BD413	4.00×10^7	1.50×10^4	3.75×10^{-4}
	9.50×10^7	5.00×10^4	5.26×10^{-4}
	1.50×10^8	4.50×10^4	3.00×10^{-4}
BD413	5.00×10^6	1.28×10^2	2.56×10^{-5}
	2.67×10^7	5.00×10^2	1.66×10^{-5}
	1.83×10^7	1.83×10^2	9.18×10^{-6}
None	2.83×10^7	8	2.83×10^{-7}
	1.10×10^7	2	1.82×10^{-7}
	2.33×10^7	4	1.71×10^{-7}

The mean transformation frequencies of HGW1521(pQM17) to prototrophy, *in situ* by lysates or untreated cultures of BD413 were 4.00×10^{-4} and 1.71×10^{-5} per recipient respectively. One-way analysis of variance showed that these were significantly different ($P < 0.001$) from both each other and from the mean frequency when no source of DNA was added. The variability obtained in this experiment was small. Calculation of the minimum significant difference (Fry 1989; MSD = 0.407 for $\log_{10}$ transformed data) showed that differences could be resolved when transformation frequencies were about 2.5 fold different. Colonies formed on selective media when no source of DNA was added could be due to mutation, transfer from the natural population or growth of indigenous microorganisms. The resulting frequency (2.1×10^{-7}) represents the limit of detection for *in situ* methodology.

In these *in situ* experiments it is also essential to ensure that the observed transfer frequency was due to the conditions in the river during incubation. Transfer must not occur during transport or enumeration of transformants. This was confirmed in control experiments by mixing the source of DNA and recipients on site and immediately returning them to the laboratory in minimal salts containing DNase I. These controls never grew any transformants, so this transformation system is suitable for examining the effect of *in situ* conditions on gene transfer.

Transformation has been previously demonstrated in the presence of the indigenous population in a sediment microcosm (Stewart and Sinigalliano 1990). However, our report appears to be the first case of genetic transformation occurring *in situ*. These findings suggest that transformation could be an important mechanism by which genetic information is spread through natural populations.

References

Bale MJ, Fry JC and Day MJ (1987) Plasmid transfer between strains of *Pseudomonas aeruginosa* on membrane filters attached to river stones. J Gen Microbiol 133:3099-3107

Baumann P (1968) Isolation of *Acinetobacter* from soil and water. J Bacteriol 96:39-42

Coughter JP and Stewart GJ (1989) Genetic exchange in the environment. Antonie van Leeuwenhoek 55:15-22

Fry JC (1989) Analysis of variance and regression in aquatic bacteriology. Binary-Computing in Microbiology 1:83-88

Fry JC and Day MJ (1990) Bacterial Genetics in Natural Environments, London, Chapman and Hall

Juni E (1972) Interspecies transformation of *Acinetobacter* genetic evidence for a ubiquitous genus. J Bacteriol 112:917-931

Levey SB and Miller RV (1989) Gene Transfer in the Environment. Environmental Biotechnology Series, USA, McGraw-Hill

Rochelle PA, Day MJ and Fry JC (1988) Occurrence transfer and mobilisation in epilithic strains of *Acinetobacter* of mercury-resistance plasmids capable of transformation. J Gen Microbiol 134:2933-2941

Saye DJ, Ogunseitan OA, Sayler GS and Miller RV (1990) Transduction of linked chromosomal genes between *Pseudomonas aeruginosa* strains during incubation *in situ* in a fresh water habitat. Appl Environ Microbiol 56:140-145

Stewart GJ and Sinigalliano CD (1990) Detection of horizontal gene transfer by natural transformation in native and introduced species of bacteria in marine and synthetic sediments. Appl Environ Microbiol 56:1818-1824

Section 3

SOIL and TERRESTRIAL ENVIRONMENTS

Conjugal Gene Transfer in the Soil Environment; New Approaches and Developments.

E. Smit and J.D. Van Elsas
Institute for Soil Fertility Research
Wageningen Branch
P.O. Box 48
6700 AA, Wageningen
The Netherlands

Introduction

Gene transfer in the environment has been identified as a process which potentially enhances putative hazards of released genetically engineered microorganisms (GEMs). This contentment has spurred recent research in this area. Our understanding of how the environment affects genetic interactions between micro-organisms has therefore increased drastically in recent years. It is now evident that the gene transfer mechanisms classically known to take place *in vitro* in the laboratory, i.e. transformation, transduction and conjugation, can also occur in nature given conditions are conducive to these processes (e.g. Stewart and Sinigalliano 1990; Zeph et al. 1988; Van Elsas et al. 1988).

Conjugation is now known as a major process potentially responsible for the dissemination of genes among natural microbial communities. The involvement of conjugation in the spread of plasmid-borne traits has been well-documented via direct studies in various different natural systems, viz. soil (Henschke and Schmidt 1990; Krasovsky and Stotzky 1987; Smit et al. 1991), the rhizosphere (Van Elsas et al. 1988; Smit et al. 1991), inside hazelnut tissue (Manceau et al. 1986), in nodules (Pretorius-Güth et al. 1990), in epilithon (Fry and Day 1990) and in other aquatic systems such as activated sludge (McClure et al. 1990) or freshwater (Fulthorpe and Wyndham 1991). Much of the current knowledge on environmental gene transfer has been obtained in experiments in which transfer of a plasmid from an introduced donor to an co-introduced genetically marked recipient strain was studied (e.g. Van Elsas et al. 1987, 1988; Richaume et al. 1990; Top et al. 1990; Klingmüller 1991). These experiments have provided insight in the influence of the environment on bacterial mating processes. However, they have not been *a priori* very predictive of the putative fate of introduced genetic material, i.e. concerning the transfer to the numerous different species present in the indigenous microbial community.

In addition, most of these experiments have been focused on the transfer of genes inserted in (often self-transmissible) plasmids, whereas transfer of chromosomal sequences has been rarely studied. Nowadays, questions arise concerning the possible role of naturally occurring genetic elements which can mobilize plasmids from introduced GEMs (Hill et al. 1992), or which might even recruit chromosomal sequences.

Moreover, the scope of conjugal transfer processes, i.e. the potential for natural transfer between Gram-positive and Gram-negative species (Gormley and Davies 1991; Bertram et al. 1991), and the flexibility of the conjugal gene transfer process, e.g. the potential for the occurrence of retro-mobilization (Mergeay et al. 1987) have recently been shown to be higher than previously expected. The environmental significance of these processes is only beginning to be explored now.

This paper will address some recent developments and approaches in environmental conjugal gene transfer, with a special emphasis on soil. After discussing the need for methodology development and describing our recent approaches, results of some attempts to describe mobilization and retromobilization of plasmids in soil will be focused on.

Use of marker genes for tracking genetically modified bacteria and their DNA in the soil ecosystem.

To study the fate of genetically engineered microorganisms (GEMs) and their DNA in the soil ecosystem, it is necessary to be able to specifically detect these bacteria and their genes, and to distinguish them from the natural soil population. One of the purposes of our current gene transfer studies in soil has been to study the effect of the localization of a selectable marker in the bacterial genome, i.e. on a self-transmissible broad-host-range plasmid like those of the IncP1 group, on a mobilizable broad-host-range plasmid like IncQ, or inserted into the chromosome, on its mobility. Therefore, plasmid RP4p (Van Elsas et al. 1991) has been previously used as a marked representative of an IncP1 plasmid. In this paper, the construction and use in soil studies of a marked IncQ plasmid and of a chromosomally inserted marker are presented. For that purpose, a universal marker gene cassette was constructed. Main criterion for choice of marker genes was a low probability of occurrence of its phenotype and/or genotype in soil. Antibiotic resistance genes were chosen as prime candidates to make part of the marker cassette, for their capacity to provide the required selectability. A previous study of background resistance to several antibiotics revealed relatively high numbers of bacteria resistant to several antiboitics in the Ede loamy sand soil used in our laboratory. Levels of kanamycin resistant cfu were about 10^4 to 10^5 per gram of dry soil, whereas cfu resistant to a combination of kanamycin and gentamycin occurred at levels of, at most, 10^3 per gram dry soil. Therefore, we opted for the combined use of a gene conferring resistance to kanamycin, *npt*II (Simon et al. 1983), and a gene conferring resistance to gentamycin, *aad*B (Schmidt et al. 1988) in the cassette. The marker cassette should also contain a DNA sequence which is not expressed and which is not commonly found in soil bacteria in combination with the resistance genes. This sequence can be used for hybridisation and PCR purposes. As such, part of the *cry*IVB gene, coding for a delta endotoxin, from *Bacillus thuringiensis* var. *morrisoni* (Waalwijk et al. 1991), was selected. As will be described in the following, the marker gene cassette was cloned into the broad-host-range (Inc Q) plasmid pSUP104 (Priefer et al. 1985) which can be mobilized into different (soil) bacterial species, and into a disarmed transposon delivery vector (Herrero et al. 1990) to facilitate chromosomal insertion. Both methods were used to mark *Pseudomonas fluorescens* R2f , originally isolated from soil (Van Elsas et al. 1988), to study marker stability, recovery from soil, genetic transfer and expression of the genes in other soil bacteria, and PCR-mediated detection of the cassette from soil DNA extractions.

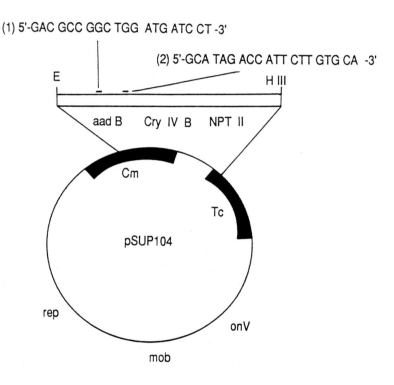

Fig. 1. Map of the plasmid pSKTG, the marker cassette inserted in HindIII-EcoRI digested pSUP104 and the primers for PCR.

Introduction of a marker gene cassette into host organisms

Strains used
Escherichia coli JM101 with vector plasmid pUC19 was used for the cloning work. The *npt*II gene was obtained from plasmid pSUP2021 (Simon et al. 1983). The *cry*IVB gene was kindly provided by Dr. C. Waalwijk (Waalwijk et al. 1991), and the *aad*B gene (in pFL1017) was a gift from Dr. F.R.J. Schmidt (Schmidt et al. 1988). Plasmid pSUP104 was obtained from the Phabagen Collection, Utrecht, the Netherlands. The disarmed transposon delivery system (Herrero et al. 1990) was provided by Dr. B. Zaat, Leiden, the Netherlands. All strains were routinely cultured in LB medium or on LB agar supplemented with the appropriate antibiotics. The *E. coli* S17.1 mobilizer strain which carries *tra* functions on the chromosome was used to mobilize the marked plasmid pSKTG to different species of soil bacteria. *Pseudomonas fluorescens* R2f wild-type and derivatives were used for soil microcosm studies.

All molecular techniques were performed according to Maniatis et al. (1989).

Construction of the marker gene cassette and its insertion into an IncQ plasmid
The *npt*II gene was obtained from Tn*5* by HindIII-SalI digestion of pSUP2021 and the resulting 1.5 kb fragment was cloned into pUC19 to give pES1. Subsequently, a 1.8 kb XbaI-XbaI fragment from the *cry*IVB gene was inserted, adjacent to the *npt*II-containing fragment, into pES1, to give pES2. Then, the central 0.47 kb PstI fragment was deleted from pES2, eliminating restriction sites interfering with further cloning and part of the *cry*IVB gene. This resulted in plasmid pESC1. The 2.9 kb HindIII-BamHI fragment from pESC1, containing both *npt*II and the truncated *cry*IVB sequence, was then inserted into the RSF1010 derivative pFL1017 (Schmidt et al. 1988), next to the *aad*B gene, giving plasmid pFLm. However, as described further, this rendered the RSF1010-derived plasmid very unstable in *Pseudomonas fluorescens* R2f. In order to obtain a stable plasmid, the 4 kb HindIII-EcoRI fragment from pFLm, containing *npt*II, *cry*IVB (truncated, non-expressed) and *aad*B, was inserted into

HindIII-EcoRI linearized pSUP104 (an RSF1010-derived vector constructed by Priefer et al. 1985). This final construct, depicted in Fig. 1, was designated pSKTG.

Plasmid pSKTG proved to be stable in *P. fluorescens* and both resistances were expressed, since normal growth was observed on LB agar with 100 mg/l of both kanamycin and gentamycin. Digestion of plasmid DNA with several restriction enzymes also confirmed the presence of the marker cassette (not shown) in the plasmid, and all colonies on selective agar plates hybridized, in a colony hybridization assay, to the *cry*IVB probe.

Insertion of the marker gene cassette into the P. fluorescens chromosome

The strategy for chromosomal insertion of the marker cassette was based on the disarmed transposon delivery system of Herrero et al. (1990). The marker cassette was isolated from pSKTG by HindIII-EcoRI digestion and inserted into helper plasmid p18Sfi (Herrero et al, 1990). The cassette could then be retrieved by SfiI digestion and ligated into pUT/Hg from which the Hg^r gene had been removed by SfiI digestion. The resulting plasmid, pUT/KTG (Fig. 2), was then transformed into the mobilizer strain *E. coli* SM10 lambda pir (Herrero et al, 1990). Plasmid pUT/KTG cannot replicate in bacteria other then *E. coli* lambda pir. Thus, to obtain bacteria with the cassette inserted into the chromosome, filter matings were performed with *E. coli* lambda pir (pUT/KTG) as the donor and a rifampicin resistant mutant of *P. fluorescens* R2f as the recipient (Smit et al. 1991). Insertion mutants, designated R2fmc were selected on LB agar supplemented with kanamycin and gentamycin (both 50 mg/l).

Insertion of the cassette into the chromosome of *P. fluorescens* by mating of this strain with *E. coli* SM10 lambda pir(pUT/KTG) occurred with a frequency of 10^{-5}. Southern blotting of genomic DNA digested with EcoRI and HindIII with a *cry*IVB probe (Fig. 2) confirmed the insertion. For comparison, a blot of EcoRI and HindIII-EcoRI digested pUT/KTG is given next to similar digestions of genomic DNA of 1 R2fmc strain (Fig. 2). The EcoRI-HindIII cassette sequence was in both cases 4 kb in size (Fig. 2, lanes 1 and

3), since both restriction sites are present in the cassette. Digestion with only EcoRI gives a band of similar size in pUT/KTG (Fig. 2, lane 2) due to the presence of an EcoRI site just outside the cassette in the plasmid sequence (Fig. 2, map), whereas it gives a 7-kb band with genomic DNA, thus suggesting that vector sequences are no longer present with the insert.

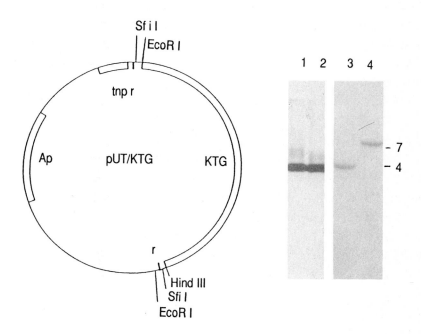

Fig. 2. Left, restriction map of pUT/KTG with some relevant restriction sites, tnp is transposase gene, KTG is the marker cassette, r are the inverted repeats of Tn5; Right, Southern blot of pUT/KTG (Lane 1 and 2) digested with EcoRI-HindIII (lane 1) and EcoRI (lane 2) and genomic DNA of R2fmc digested with EcoRI-HindIII (lane 3) and EcoRI (lane 4); band sizes in kb are given on the right.

Stability of the marker gene cassette and expression in other species
To check if the marked plasmid, pSKTG, and the chromosomal KTG insertion were stable in *P. fluorescens* R2f, strains carrying the plasmid and the cassette were cultured in LB broth without antibiotics and plated on selective and non-selective plates. Strains were also introduced into Ede loamy sand and maintained for 7 days

at 15ºC. At regular intervals, dilutions were plated on LB agar with or without kanamycin and gentamycin (50mg/l). Plasmid pSKTG revealed to be fully stable in *P. fluorescens* R2f after overnight culturing in LB (approx. 10 generations), whereas the previous construct in pFL1017, pFLm, was unstable and showed 99% plasmid loss in 10 generations.

Also, the chromosomally inserted KTG sequence was 100% stable after culturing (approximately 10 generations). During presence in Ede loamy sand soil for 7 days, we could not find any evidence for loss of plasmid pSKTG or of the KTG sequence inserted into the chromosome (using rifampicin resistant mutants), by comparing cfu counts on agar plates with rifampicin with plates with kanamycin and gentamycin, and probing with the *cry*IVB probe.

Plasmid pSKTG was transformed into the mobilizer strain *E. coli* S17.1, giving *E. coli* S17.1 (pSKTG). Then, filter matings were performed between this strain and several different rifampicin resistant species of soil bacteria to check if the antibiotic resistance genes would be expressed in a wide species range. Whereas *npt*II is known to show broad-host-range expression, less information is available on the expression range of *aad*B. The expression studies

Table 1. Mobilization of pSKTG and expression of the resistance genes in several different bacterial species. Km = kanamycin 50 mg/l; Gm = gentamycin 50 mg/l; ISFR= Institute for Soil Fertility Research; UL= A. Richaume, University of Lyon, France; PD= Plant Pathology Service, Wageningen, the Netherlands.

Strain	Source	Growth on Km Gm
Escherichia coli	ISFR	+
Enterobacter cloacae	ISFR	+
Agrobacterium tumefaciens	UL	+
Pseudomonas fluorescens R2f	ISFR	+
Pseudomonas fluorescens GE1	ISFR	+
Pseudomonas putida	ISFR	+
Pseudomonas cepacia	ISFR	+
Xanthomonas maltophilia	PD	+

were deemed necessary in order to check (i) whether the marker cassette can be used for species other than *P. fluorescens*, and (ii) whether transfer of the cassette to other species can be detected by selection on kanamycin and gentamycin. Table 1 shows that in the species tested which belonged to a wide variety of Gram-negative species, *aad*B was expressed (as was *npt*II), and the recombinant plasmid was maintained.

Detection of marked bacteria in soil by selective plating

To check recovery from soil, different bacterial cell numbers of *Pseudomonas fluorescens* R2f (pSKTG) and R2fmc were added to microcosms containing Ede loamy sand soil according to Smit et al. (1991). Cfu were enumerated by plating on King's B agar with gentamycin and kanamycin (50 mg/l) before and 3 hours after addition to soil (Fig. 3). Identical Log numbers of cells added and Log numbers of cells recovered, resulting in lines with an angle of 45° (Fig. 3), indicated 100 % recovery of both marked strains from soil. no short- term loss of expression due to the introduction into soil apparently occurred.

Detection of marked cells in soil by DNA extraction and PCR

The soil portions described above were also used to extract DNA for PCR-mediated detection of the marker cassette of the introduced bacteria. DNA was extracted from soil and purified as described by Smalla et al. (1992). Additional DNA purification was used since no amplification could be achieved initially. Thus, after purification using the CsCl, KAc and spermine-HCl precipitation steps, DNA pellets were dissolved in 200 µl of TE buffer. Final purification was as follows. CTAB (20 µl) and 5M NaCl (20 µl) were added to 100 µl of DNA (Ausubel et al. 1987), and the samples were incubated at 65°C for 10 min. Samples were then extracted twice with phenol/ chloroform-isoamylalcohol after which the DNA was precipitated with 3 Vol ethanol and taken up in 90 µl of TE. These solutions were subsequently cleaned with Geneclean II glass milk (Bio 101 Inc., La Jolla, CA, USA). 1 µl was used for amplification using SuperTaq polymerase in a 50 µl reaction mix according to Smalla et al. (1992). Primers used are given in Fig.1. Figure 3 (right) shows the results of the amplification; lanes correspond to the different soil portions with increasing bacterial numbers as indicated in the graph (lane 1 received no cells, lane 2 received Log 2.8, etc). The expected band of 411 bp is identical to the product found after amplification of pure target DNA. The expected band is clearly visible in reaction mixtures of PCR run on soil with the highest inoculum densities, whereas a band is just detectable in the sample with the lowest cell numbers (Log 2.8 per g of dry soil).

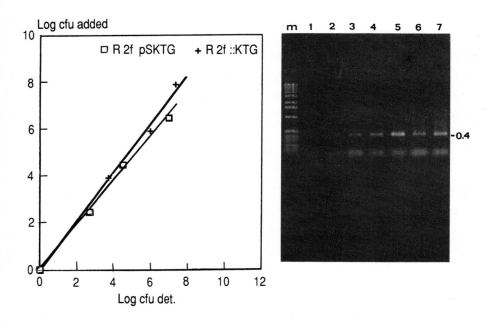

Fig. 3. Left: Log number of cfu per gram of dry soil added to soil against the Log number of cfu recovered on selective plates of *P. fluorescens* R2fmc (square) and R2f (pSKTG) (+); Right: agarose gel of PCR amplified DNA from the samples given on the left.

Mobilization of pSKTG in soil and rhizosphere

Mobilization of pSKTG into indigenous microbes was studied in Ede loamy sand soil using 2 different approaches. In the first approach, a mobilizing plasmid, RP4p, was co-introduced, either in the same host or in a different co-inoculated strain. The second approach focused on the capacity of the natural microbial community present in soil and rhizosphere to mobilize pSKTG. In this case we did not use a recipient like in Hill et al. (1992), but the indigenous bacterial population was screened for the presence of pSKTG, which they potentially received by tri-parental mating or retro-transfer. Both approaches were performed in soil and on filters. To address mobilization to indigenous microorganisms, donor counterselection using phage ϕR2f was used in accordance with Smit et al. (1991). Results of both approaches are summarized in Table 2. Results of the first approach clearly showed enhanced mobilization in rhizosphere soil of pSKTG in the presence of RP4p in the same host, and a 100-fold reduction of the transfer rate with RP4p present in a different host. A full account of these data is currently in preparation. So far, the second approach has not provided any conclusive data as to the degree of pSKTG mobilization by genetic elements naturally occurring in microbial populations in the rhizosphere of wheat as was found by Hill et al. (1992) in the epilithon. In several studies in 2 different soils, Ede loamy sand and Flevo silt loam, no 'natural' mobilization was found, since indigenous transconjugants harbouring pSKTG could not be recovered. Since 'natural' mobilization might be a very rare event, it will probably be virtually undetectable in *in situ* soil studies, but it might be observed in filter matings with the total soil bacterial population. However, while the background resistant population was not a problem in soil studies, in filter matings with the total soil bacterial population, the presence of a relative large resistant population after the incubation, enhanced the detection limit and thus prevented detection of potential low numbers of transconjugants. The use of a recipient in this approach might prove to be more successful, since extra selection is then provided.

Table 2. Mobilization of pSKTG from *P. fluorescens* R2f (Donor) to indigenous bacteria in microcosms with Ede loamy sand planted to wheat (S) and on filters (F) in the presence and absence of RP4p.

Treat-ment		Log cfu/g soil Donor			Indigenous
		RP4p	pSKTG	both	pSKTG
Sa	R	6.3	6.5	-	2.1
	B	6.4	6.4	-	2.0
Sb	R	-	6.5	-	[0]
	B	-	6.4	-	[0]
Sc	R	-	-	5.0	4.2
	B	-	-	4.1	2.4
F1		9.4	9.2	-	4.0
F2		-	8.0	-	[0]

For Sa Sb and Sc, cfu counts per gram dry soil after 7 days are presented. Sa: a donor with RP4p and a donor with pSKTG were separately inoculated; Sb: a donor with only pSKTG was inoculated; Sc: a donor with both RP4p and pSKTG was added. Initial inoculum levels (per g dry soil) were around Log 7. Filter matings with the total soil bacterial population were performed (F), with (F1) and without (F2) the presence of a mobilizer strain (donor RP4p).
[0] = below detection limit; R = rhizosphere, B = bulk soil.

Concluding remarks

It is becoming increasingly clear that conjugal gene transfer in the soil environment is a common process, albeit occurring at often low frequencies. Experimental approaches aimed at tackling transfer from introduced donor bacteria into the indigenous populations so far have struggled with difficulties due to these low frequencies. Two main reasons for this may be identified, the first one being the (phenotypic) background in many soil microbial communities of many marker genes, and the second one being the difficulty of obtaining low numbers of indigenous transconjugants against high numbers of introduced donor cells. The problem of background bacterial populations obscuring the appearance of indigenous transconjugants in soil studies can be overcome using highly specific combinations of marker genes and sequences, as shown by Van Elsas et al (1991) and argued in this paper. Drastic reduction of the background was possible when using selection for the marker genes followed by hybridization to the specific, non-selectable, sequence. Moreover, donor counterselection in soil-derived communities can be achieved using different approaches. Thus, Henschke and Schmidt (1990) were only able to detect transfer to indigenous microbes following die-out of the introduced *E. coli* donor strain, and Smit et al. (1991) used a specific bacteriophage to counterselect the donor. In addition, Top et al. (1990) used a so-called gene escape system, in which *czc* marker genes present on a self-transmissible plasmid were unexpressed in the used donor strain, and therefore donor growth on selective agar plates was reduced.

The further development of probing and donor counterselection techniques should permit the direct detection of the potentially rare transfer events alluded to in the Introduction, i.e. heterogramic transfer and retromobilization.

Next to gene transfer studies via the direct approach, the spread of genes in the environment can also be studied retrospectively, i.e. the appearance of similar genetic elements in different strains following a categorization of these elements in the soil microbial community might be indicative of recent horizontal gene transfer. Criteria for such an analysis were given by Chater (1988). Retrospective evidence of gene transfer may better describe the effect of selective pressure acting on the products of mating events in soil, and may as such be meaningful to ascertain ecological long-term effects. We are currently in the process of testing the effect of deliberately applied antibiotic selective pressure on the fate of these antibiotic resistance marker genes in soil.

Acknowledgements

E. Smit was supported by the Dutch National Programme for Soil Science.

References

Ausubel FM, Brent R, Kingston RE, Moore DD, Seidman JG, Smith JA, Struhl K (1987) Current protocols in molecular biology. John Wiley & Sons, New York.

Bertram J, Strätz M, Dürre P (1991) Natural transfer of conjugative transposon Tn916 between Gram-positive and Gram-negative bacteria. J Bacteriol 173:443-448.

Chater KF, Henderson DJ, Bibb MJ, Hopwood DA (1988) Genome flux in *Streptomyces coelicolor* and other streptomycetes and its possible relevance to the evolution of mobile antibiotic resistance determinants. pp. 7-42. In Transposition (Kingsman AJ, Chater KF, Kingsman SM, Eds.) Cambridge University Press, Cambridge.

Fulthorpe RR, Wyndham RC (1991) transfer and expression of the catabolic plasmid pBR60 in wild bacterial recipients in a freshwater ecosystem. Appl Environ Microbiol 57:1546-1553.

Fry JC, Day MJ (1990) Plasmid transfer in the epilithon. pp. 55-80. In Bacterial Genetics in Natural Environments. (Fry JC, Day MJ, Eds) Chapman and Hall, London.

Gormley EP, Davies J (1991) Transfer of plasmid RSF1010 by conjugation from *Escherichia coli* to *Streptomyces lividans* and *Micobacterium smegmatis*. J Bacteriol 173:6705-6708.

Henschke RB, Schmidt FRJ (1990) Plasmid mobilization from genetically engineered bacteria to members of the indigenous soil microflora in situ. Curr Microbiol 20:105-110.

Herrero M, De Lorenzo V, Timmis KN (1990) Transposon vectors containing non-antibiotic resistance selection markers for cloning and stable chromosomal insertion of foreign genes in Gram-negative bacteria. Appl Environ Microbiol 172:6557-6567.

Hill KE, Weightman AJ, Fry JC (1992) Isolation and screening of plasmids from the epilithon which mobilize recombinant plasmid pD10. Appl Environ Microbiol 58:1292-1300.

Klingmüller W (1991) Plasmid transfer in natural soil: a case by case study with nitrogen-fixing *Enterobacter*. FEMS Microbiol Ecol 85:107-116.

Krasovsky VN, Stotzky G (1987) Conjugation and genetic recombination in *Escherichia coli* in sterile and non-sterile soil. Soil Biol Biochem 19:631-638.

Manceau C, Gardan L, Devaux M (1986) Dynamics of RP4 transfer between *Xanthomonas campestris* pv. *carolina* and *Erwinia herbicola* in hazelnut tissues, in planta. Can J Microbiol 32:835-841.

Maniatis T, Fritsch EF, Sambrook AJ (1989) Molecular cloning: a laboratory manual, 2nd edition. Cold Spring Harbor Laboratory, Cold Spring Harbor, N.Y.

McClure, NC, Fry JC, Weightman AJ (1990) Gene transfer in activated sludge, pp. 111-129. In Bacterial genetics in natural environments (Fry JC, Day MJ, Ed.) Chapman and Hall, London.

Mergeay M, Lejeune P, Sadouk A, Gerits J, Fabry L (1987) Shuttle transfer (or retrotransfer) of chromosomal markers mediated by pULB113. Mol Gen Genet 209: 61-70.

Pretorius-Güth IM, Pühler A, Simon R (1990) Conjugal transfer of megaplasmid 2 between *Rhizobium meliloti* strains in alfalfa nodules. Appl Environ Microbiol 56: 2354-2359.

Priefer UB, Simon R, Pühler A (1985) Extension of the host range of *Escherichia coli* vectors by incorperation of RSF1010 replication and mobilisation functions. J Bacteriol 163:324-330.

Richaume A, Angle JC, Sadowsky MJ (1990) Influence of soil variables on in situ plasmid transfer from *Escherichia coli* to *Rhizobium fredii*. Appl Environ Microbiol 55:1730-1734.

Schmidt FRJ, Nücken EJ, Henschke RB (1988) Nucleotide sequence analysis of 2"-aminoglycoside nucleotidyl-transferase ANT(2") from TN4000: its relationship with aad(3") and impact on Tn21 evolution. Mol Microbiol 2:709-717.

Simon R, Priefer U, Pühler A (1983) Vector plasmids for in-vivo and in-vitro manipulations of Gram-negative bacteria. pp. 89-106. In Molecular genetics of the bacteria-plant interaction (Pühler A, ed) Springer-Verlag, Berlin.

Smalla K, Cresswell N, Mendonca-Hagler LC, Wolters A, Van Elsas JD (1992) Rapid DNA extraction protocol from soil for polymerase chain reaction mediated amplification. J Appl Bacteriol (in press).

Smit E, Van Elsas JD, Van Veen JA, De Vos WM (1991) Detection of plasmid transfer from *Pseudomonas fluorescens* to indigenous bacteria in soil by using bacteriophage φR2f for donor counterselection. Appl Environ Microbiol 57:3482-2488.

Stewart GJ, Sinigalliano CD (1990) Detection of horizontal gene transfer by natural transformation in native and introduced species of bacteria in marine and synthetic sediments. Appl Environ Microbiol 56:1818-1824.

Top E, Mergeay M, Springael D, Verstraete W (1990) Gene escape model: transfer of heavy metal resistance genes from *Escherichia coli* to *Alcaligenes eutrophus* on agar plates and in soil samples. Appl Environ Microbiol 56: 2471-2479.

Van Elsas JD, Trevors JT, Starodub ME (1988) Bacterial conjugation between pseudomonads in the rhizosphere of wheat. FEMS Microbiol Ecol 54:299-306.

Van Elsas JD, Govaert JM, Van Veen JA (1987) Transfer of plasmid pFT30 between bacilli in soil as influenced by bacterial population dynamics and soil conditions. Soil Biol Biochem 19:639-647.

Van Elsas JD, Van Overbeek LS, Fouchier R (1991) A specific marker, *pat*, for studying the fate of introduced bacteria and their DNA in soil using a combination of detection techniques. Plant and Soil 138:49-60.

Waalwijk C, Dullemans A, Maat C (1991) Construction of a bioinsecticidal rhizosphere isolate of *Pseudomonas fluorescens*. FEMS Microbiol Lett 77: 257-264.

Zeph LR, Onaga MA, Stozky G (1988) Transduction of *Escherichia coli* by bacteriophage P1 in soil. Appl Environ Microbiol 54, 190-193.

Gene Transfer via Transformation in Soil/Sediment Environments

M. G. Lorenz
Genetik, Fachbereich Biologie,
Universität Oldenburg
Postfach 2503
W-2900, Oldenburg
Germany

Genes of prokaryotes can be transferred from one cell to another by three distinct parasexual processes. Two of them are accomplished by intracellular "parasites": bacteriophages can transport genes (transduction) and plasmids can promote DNA transfer during cell contact (conjugation). The third parasexual process is genetic transformation during which free DNA is taken up by cells. This DNA uptake is controlled by bacterial genes located on the chromosome. In the past decades the physiology, biochemistry, genetics, and molecular biology of the three transfer mechanisms have been the subject of a vast number of studies. Many principles and details have been elucidated. However, as yet there is no definite answer to the question whether conjugation, transduction and transformation take place in natural bacterial habitats and, if so, what the effects are on the dynamics of natural bacterial populations.

In the past several years the question of natural gene transfer in the environment has been experimentally approached by microcosm studies. These studies have mainly concentrated on the conjugative spread of plasmids in aquatic and soil environments. This rather limited focus may partially result from the consideration that released genetically engineered microorganisms will possess their specific genetic load on plasmids. More recently there is increasing interest to study also the other gene exchange mechanisms because besides chromosomal DNA, plasmids can be transferred by transduction and transformation. Due to safety considerations with respect to the conjugative plasmid spread, the bacterial chromosome is increasingly becoming a target for genetic manipulation besides plasmids.

Principally there are two ways to investigate the ecology and genetics of natural and introduced bacteria. One is the use of microcosms consisting of environmental samples having the advantage of simulating natural habitats as close as possible. A major drawback of this strategy is the multiplicity of mostly unknown factors which may affect the experiment such as the chemical potential in the sample, the physical properties of the complex matrix and the biological parameters including the unknown composition and density of the microbial population and the presence and activity of enzymes. The second way uses models of the environment. Such models can be simple with respect to the composition of solid and liquid phases, the number and diversity of microorganisms and so forth. The complexity of such model systems can be increased stepwise to approach natural environments.

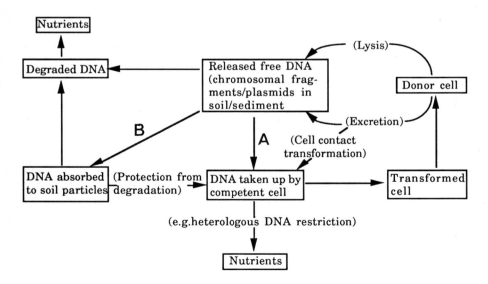

Fig.1. Scheme of gene flow in soil/sediments by genetic transformation

By the comparison of data from model systems with those from microcosms including material sampled from natural environments it is expected to identify and analyze general abiotic and biotic parameters and processes that affect the survival and gene transfer of microorganisms. This approach is particularly useful for soil and sediments because of the high complexity of these environments. Altogether, a comprehensive understanding of how bacteria behave in natural habitats should be attained by both approximations, but the ultimate goal should be investigations *in situ*.

In our laboratory, we have focused on the bacterial gene transfer by free DNA in the environment which includes the study of genetic transformation in soil and sedimentary habitats. Our analytical strategy relies on the study of several Gram-positive and Gram-negative species of soil bacteria and the use of microcosms with model soils (chemically pure sea sand, enriched specific mineral fractions, clays and sand/clay mixtures) and environmental samples (groundwater aquifer material, various soils, groundwater). Further, we have dissected the process of transformation in the environment in a series of distinct steps (Wackernagel et al. 1992). In model microcosms, the natural factors that might be important for transformation of bacteria in the environment are being elaborated. Based on what is known about the mechanisms of transformation and about properties of soils and what has been learnt from studies of the various microcosm systems, a hypothetical flow of genes in soil/sediment environments is put forward (Fig. 1).

In this scheme, genetic material passes from a donor to a recipient cell on two alternative routes, A and B. In a direct way (Fig. 1, A), DNA is transferred by release from a donor cell into the surrounding milieu and is subsequently taken up by a competent cell or during cell-contact transformation (Orrego et al. 1978; Stewart et al. 1983; Rochelle et al. 1988). DNA release can be the result of cell lysis or excretion and comprises chromosomal fragments (Borenstein and Ephrati-Elizur 1969; Sinha and Iyer 1971; Lorenz et al. 1991) as well as plasmid molecules (Lorenz et al. 1991) which can transform competent recipients even in the presence of autologous extracellular DNases. The alternative way for transformation is via soil particles (Fig. 1, B). Chromosomal or plasmid DNA added to microcosms consisting of quartz sand, sand/clay mixtures or material sampled from a groundwater aquifer, rapidly adsorbed to minerals and thereby acquired high resistance against DNase (Lorenz and Wackernagel 1987; Romanowski et al. 1991; Romanowski et al. unpublished). Such adsorbed DNA can persist in the environment and can be taken up by competent cells (Lorenz et al. 1988; Lorenz and Wackernagel 1990). Also, introduced into natural soils purified plasmid DNA and its nucleotide sequences were detectable after incubation periods of days or months (Romanowski et al. submitted). These fluxes of DNA depicted in Fig. 1 may lead to a horizontal and also to a non-horizontal (post mortal) gene transfer.

In the few instances examined so far, the ionic requirements for uptake of DNA are low and are met by groundwater and soil liquid phases (Lorenz and Wackernagel 1991; Lorenz et al. 1992). A more critical parameter of transformation in soil seems to be the development of competence which requires induction and expression of specific genes and hence metabolism. Soil is a nutrient-poor environment and therefore competence development depends on the supply of bacterial cells with appropriate nutrients. Interestingly, competence of a soil bacterium was stimulated under starvation for a single nutrient (C-, N-, P-source) in soil extract (Lorenz and Wackernagel 1991). Another important factor is the heterogeneity of extracellular DNA present in soil. DNA from several sources including bacteria, eukaryotes and viruses present in the environment may or may not compete with homologous DNA for uptake. Some species of soil bacteria seem to take up specifically DNA of their own species while others take up any DNA (see Bruns et al. 1992). Furthermore, DNA-restriction may interfere with transformation. The first experiments directed towards this question did not indicate strong effects of restriction on transforming DNA (Ganesan 1982; Bruns et al. 1992). Also, a plasmid monomer entering the cytoplasm in the linear single-stranded form can not establish as a replicon because of a lack of homology to recipient DNA sequences necessary for recircularization (Canosi et al. 1981). This problem is being addressed by various studies (Bruns et al. 1992).

As yet, no answer can be given to the questions whether genetic transformation occurs in soil or aquatic environments with appreciable frequency and how the possible gene flux is distributed over bacterial habitats and their seasonal variations. However, in the relatively simple environment of aquatic habitats natural transformation has been demonstrated very recently in two cases (Stewart and Sinigalliano 1990; Paul et al. 1991) whereas the transformation in a more complex environment, the aquatic sediment, was less unambiguous (Stewart and Sinigalliano 1990; Paul et al. 1991). Except for one report (Graham and Istock 1978) no data on transformation directly in natural soil as one of the most complex habitats are as yet available. This stresses the necessity for further efforts to identify factors stimulating or inhibiting transformation in the environment. The analytical approach of employing defined microcosms, model organisms and purified DNA has turned out to provide quantitative data on the important biotic and abiotic factors and parameters. Their knowledge is required to understand the processes in the complex bacterial habitats that can contribute to the occurrence of gene flux by free DNA and to its extent.

References

Borenstein S, Ephrati-Elizur E (1969) Spontaneous release of DNA in sequential genetic order by *Bacillus subtilis*. J Mol Biol 45:137-152

Brun S, Reipschläger K, Lorenz MG, Wackernagel W (1992) Characterization of natural transformation of the soil bacteria *Pseudomonas stutzeri* and *Acinetobacter calcoaceticus* by chromosomal and plasmid DNA. Microbial Rel (this issue)

Canosi C, Iglesias A, Trautner TA (1981) Plasmid transformation in *Bacillus subtilis*: effects of insertion of *Bacillus subtilis* DNA into plasmid pC194. Mol Gen Genet 181:434-44.

Ganesan AT (1982) Uptake, restriction, modification and recombination of DNA molecules during transformation in *B. subtilis*. In Ganesan AT, Chang S, Hoch JA (eds), Molecular cloning and gene regulation in bacilli. Academic Press, New York, pp 261-268

Lorenz MG, Wackernagel W (1987) Adsorption of DNA to sand and variable degradation rates of adsorbed DNA. Appl Environ Microbiol 53:2948-2952

Lorenz MG, Wackernagel W (1990) Natural genetic transformation of *Pseudomonas stutzeri* by sand-adsorbed DNA. Arch Microbiol 154:380-385

Lorenz MG, Wackernagel W (1991) High frequency of natural genetic transformation of *Pseudomonas stutzeri* in soil extract supplemented with a carbon/energy and phosphorus source. Appl Environ Microbiol 57:1246-1251

Lorenz MG, Aardema BW, Wackernagel W (1988) Highly efficient genetic transformation of *Bacillus subtilis* attached to sand grains. J Gen Microbiol 134:107-112

Lorenz MG, Gerjets D, Wackernagel W (1991) Release of transforming plasmid and chromosomal DNA from two cultured soil bacteria. Arch Microbiol 156:319-326

Lorenz MG, Reipschläger K, Wackernagel W (1992) Plasmid transformation of naturally competent *Acinetobacter calcoaceticus* in non-sterile soil extract and groundwater. Arch Microbiol 157 (in press)

Orrego C, Arnaud M, Halvorson HO (1978) *Bacillus subtilis* 168 genetic transformation mediated by outgrowing spores: necessity for cell contact. J Bacteriol 134:973-981

Paul JH, Frischer ME, Thurmond JM (1991) Gene transfer in marine water column and sediment microcosms by natural plasmid transformation. Appl Environ. Microbiol 57:1509-1515

Rochelle PA, Day MJ, Fry JC (1988) Occurrence, transfer and mobilisation in epilithic strains of *Acinetobacter* of mercury resistance plasmids capable of transformation. J Gen Microbiol 134:2933-2941

Romanowski G, Lorenz MG, Wackernagel W (1991) Adsorption of plasmid DNA to mineral surfaces and protection against DNaseI. Appl Environ Microbiol 57:1057-1061

Stewart GJ, Sinigalliano CD (1990) Detection of horizontal gene transfer by natural transformation in native and introduced species of bacteria in marine and synthetic sediments. Appl Environ. Microbiol 56:1818-1824

Stewart GJ, Carlson CA, Ingraham JL (1983) Evidence for an active role of donor cells in natural transformation of *Pseudomonas stutzeri*. J Bacteriol 156:30-35

Wackernagel W, Romanowski G, Lorenz MG (1992) Studies on gene flux by free bacterial DNA in soil, sediment and groundwater aquifer. In Stewart-Tull DES, Sussman M (eds), The release of genetically engineered microorganisms. Plenum Publishing Corporation, New York (in press)

DNA Binding to Various Clay Minerals and Retarded Enzymatic Degradation of DNA in a Sand/Clay Microcosm

M. G. Lorenz and W. Wackernagel
Genetik, Fachbereich Biologie
Universität Oldenburg
Postfach 2503
W-2900, Oldenburg
Germany

Introduction

Clay minerals are important constituents of the soil solid matrix. They are known for their high capacity of binding biopolymers including proteins and nucleic acids (for review see Stotzky 1986). In view of the evaluation of an environmental gene transfer by genetic transformation it is important to examine the interaction of free DNA with soil particles. It is our concept to start the examination by using microcosms with model soil and then gradually build up more complex systems ending with natural soil. Recently we have characterized the adsorption to sand of linear and circular DNA and its retarded degradation by DNaseI (Lorenz and Wackernagel 1987; Romanowski et al. 1991). Here we extend the studies to various clay minerals and describe the high resistance of adsorbed DNA to enzymatic degradation in a sand/clay microcosm.

Materials and Methods

Clays and DNA.
Natural sodium-bentonite (Greenbond; NaB) and the 'activated' calcium-bentonite (BW100; CaB) were purchased from Süd-Chemie

AG (München, Germany) and the kaolinite (Britefil 80; Ka) from Albion Kaolin Co. (Hephziban, Georgia, USA). The montmorillonite contents of the NaB is about 75% and of the CaB 60 to 65%.

Calf thymus DNA (Boehringer, Mannheim, Germany) was used for adsorption studies. For the sand/clay microcosm studies [^{3}H]-thymidine labeled phage P22 DNA (2540 cpm μg^{-1}) was isolated as described (Carter and Radding 1971).

Adsorption studies.

DNA solutions (usually 1.025 mg ml^{-1}) and clay suspensions (35 mg ml^{-1}) were dialyzed over night at 20°C against 100 vol of 10 mM Tris-HCl (usually pH 7.0) and the appropriate salt (NaCl or MgCl$_2$ at concentrations as indicated for each experiment). If the volume of the solution was reduced it was adjusted by the addition of the appropriate salt solution. DNA solution (0.1 ml) was mixed with 0.1 ml of the clay suspension in an Eppendorf centrifuge cap and incubated at 20°C (usually 2 h) with occasional stirring. Then the mixture was centrifuged in an Eppendorf centrifuge at 13.000 rpm for 1 min. The supernatant was diluted 10 fold in the same salt solution used for adsorption. The DNA concentration was determined by measuring the absorbance at 258 nm (E$_{258}$ of 1.0 corresponds to 50 μg DNA ml^{-1}). The amount of DNA adsorbed was calculated as the difference between the DNA concentration before and after incubation with clay minerals.

Microcosms.

The sand/Na-bentonite microcosms were prepared as follows. The clay suspension (1 ml of 17.5 mg ml^{-1} in 5 mM MgCl$_2$, 10 mM Tris-HCl, pH 7.1) was filled into a 5 x 70 mm column (Lorenz and Wackernagel 1988) and 0.7 g of chemically pure sand (Merck, Darmstadt, Germany) was added. After settling of the sand (within 3 to 5 min) the clay-containing supernatant was removed by a Pasteur pipette. After 15 min at 20°C the column was washed with 5 ml of 5 mM MgCl$_2$, 10 mM Tris-HCl (pH 7.1) at 0.2 ml min^{-1} to remove non-adsorbed clay particles. At this stage the microcosm was ready for loading with DNA and DNaseI.

For the determination of the amount of clay present in the

microcosm, the contents of the column was removed and vigorously vortexed in 5 ml of 5 mM $MgCl_2$, 10 mM Tris-HCl (pH 7.1) for 15 s. The clay suspension (supernatant) was diluted in the same buffer and the clay concentration was determined (also in the effluent of the column) as described below.

Determination of the clay concentration in suspensions.
The assay relies on the binding of methyleneblue to clay minerals and the determination of the decolorization of the solution after removal of the clay. A clay suspension (0.7 ml) was mixed with 0.7 ml of 0.00375% (w/v) methyleneblue solution and incubated for 5 min. After centrifugation in an Eppendorf centrifuge (5 min) the extinction of the supernatant was measured at 664 nm. A linear regression of the absorbance to increasing clay concentrations was observed in the range 0.05 to 0.15 mg Na-bentonite ml^{-1}.

DNA loading, DNaseI treatment and DNA extraction.
The sand/clay microcosm was loaded with 0.2 ml of 39.7 μg [^{3}H]-labeled P22 DNA ml^{-1} in 5 mM $MgCl_2$, 10 mM Tris-HCl (pH 7.1). After 2 h at 20°C the column was eluted with the same buffer (5 ml at 0.2 ml min^{-1}). The amount of DNA bound to the material in the column was calculated as the difference of the radioactivity applied and determined in the effluent. The column was further loaded with 0.2 ml of DNaseI (100 or 500 μg ml^{-1} in 5 mM $MgCl_2$, 10 mM Tris-HCl, pH 7.1). After 1 h or 24 h at 20°C the content of the column was transferred to a tube containing 0.8 ml of the buffer. To differentiate between the radioactivity associated with sand or clay during the following DNA extraction steps, the tube was vortexed (15 s) and 0.5 ml of the supernatant was removed. The tube containing the sand plus clay and the tube with only clay (half the amount originally present in the column) were centrifuged for 2 min and the supernatant of this extract was removed. To the clay pellet 1 ml of extraction buffer (10 mM Na_2HPO_4, 20 mM EDTA, 20 mM Tris-HCl, pH 8.6) was added and to the sand/clay sediment (containing 0.2 ml interstitial volume) 0.8 ml of 1.25 fold concentrated extraction buffer. After resuspension of the clay in the two tubes and 15 min incubation, the suspensions were centrifuged

again and the supernatants removed. The procedure was repeated twice as described, except that the sand/clay sediment was vortexed in 0.8 ml of extraction buffer. The extracts were analyzed for total DNA and acid soluble material (see below).

Determination of DNA degradation products.
A sample (0.5 ml) of the various extraction steps was mixed with 0.5 ml of 12% (w/v) trichloroacetic acid and the total radioactivity was determined in 0.5 ml of this mixture in a liquid scintillation counter. For the determination of acid-soluble degradation products, 0.1 ml of bovine serum albumine (25 mg ml^{-1}) was added to the remaining solution and chilled on ice for 5 min. After 4 min centrifugation radioactivity was determined in the supernatant.

Results and Discussion

Kinetics of adsorption.
In the presence of low and high cation concentrations, rapid adsorption of DNA was observed with all three clay minerals (Fig. 1). The amount adsorbed increased with time. Within 1 to 5 min 50% of the final amount of DNA was already associated with NaB and Ka (Fig. 1A, B). The adsorption to CaB was slightly slower (Fig. 1C). The adsorption curves approximated maximum levels after 15 to 60 min. In subsequent experiments clays were incubated with DNA for 2 h to ensure complete adsorption.

Influence of cations.
With NaCl the DNA binding properties differed among the clay minerals (Fig. 2A). Increasing amounts of DNA bound to NaB and CaB with increasing NaCl concentrations. Maximum amounts of bound DNA were approached at 1 M (CaB) and 2.5 M NaCl (NaB). DNA did not bind to NaB at 0.25 M or lower concentrations of NaCl (Fig. 1A and 2A). Adsorption to Ka was fairly independent of the NaCl concentrations (Fig. 2A).

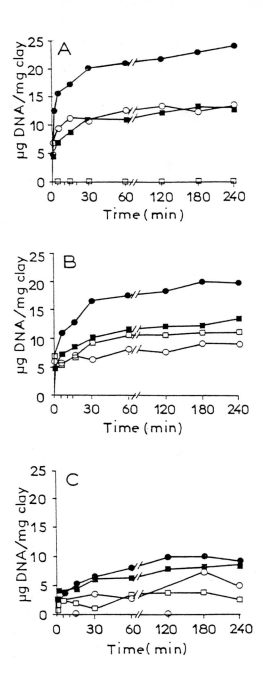

Fig. 1. Kinetics of DNA binding to Na-bentonite (A), kaolinite (B) or Ca-bentonite (C) in the presence of 0.1M NaCl (□), 5 M NaCl (■), 5 mM MgCl₂ (o) or 200 mM MgCl₂ (●)

The amount of adsorbed DNA depended on the $MgCl_2$ concentration up to 50 mM (Fig. 2B). At higher $MgCl_2$ concentrations the curves remained at clay-specific maximum levels of adsorbed DNA. With NaB and CaB a concentration of 10 mM $MgCl_2$ was similarly effective for adsorption as 2.5 M NaCl. Although less dramatic, Ka also showed an enhanced binding of DNA in the presence of $MgCl_2$.

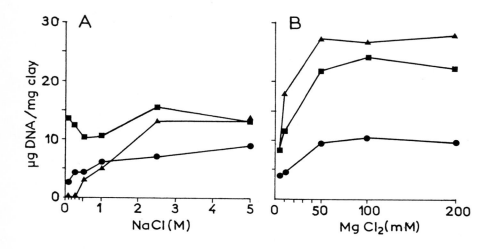

Fig. 2. Influence of NaCl (A) and $MgCl_2$ (B) concentration on the adsorption of DNA to Na-bentonite (▲), kaolinite (■) or Ca-bentonite (●)

Influence of pH.
The pH of the bulk phase did not have a substantial effect on the adsorption of DNA to the clay minerals tested (Table 1). It is noteworthy that at 0.1 M NaCl no DNA bound to NaB even at low pH. This indicates a strong dissociation of acidic groups at the surface of this clay mineral which can not be counterbalanced by the prevailing concentration of protons even at pH 5.2. In contrast, an effect of H^+ on adsorption of DNA to quartz minerals was noted (Lorenz and Wackernagel 1987).

Table 1. Adsorption of DNA to clay minerals at various pH

Clay mineral	Salt concentration in bulk	% DNA[a] adsorbed at		
		pH 5.2	pH 7.3	pH 8.9
Na-bentonite	100 mM NaCl	0	0	0
	5 mM MgCl$_2$	99.6	99.4	100
Ca-bentonite	100 mM NaCl	85.2	80.2	67.7
	5 mM MgCl$_2$	69.0	65.4	66.2
Kaolinite	100 mM NaCl	ND[b]	96.9	95.4
	5 mM MgCl$_2$	96.9	100	99.4

[a] 100% adsorption is the removal of 52 µg DNA from 1 ml solution with 17.5 mg clay ml^{-1}
[b] ND=not determined

DNA adsorption isotherms.
The quantities of DNA adsorbed to the clays were a function of the concentration of DNA and the concentration and valency of the cation in the bulk phase. Most of the curves approached maximum values (Fig. 3A, C, D), except for the DNA-adsorption to NaB and Ka in the presence of 0.2 mM MgCl$_2$ (Fig. 3B). The highest amounts of DNA associated with clays (µg DNA mg clay^{-1}) were 27.9 (NaB), 24.1 (Ka) and 10.5 (CaB).

DNA degradation in a sand/clay microcosm.
The sand/NaB microcosms were prepared as described in Materials and Methods. Their liquid phase consisted of 5 mM MgCl$_2$, 10 mM Tris-HCl (pH 7.0). In the microcosms, 4 mg of NaB remained in the sand bed (0.7 g). When the NaB for the preparation of the microcosms was suspended in distilled water no clay remained in the column, suggesting a MgCl$_2$-effected interaction of clay particles with sand grains.

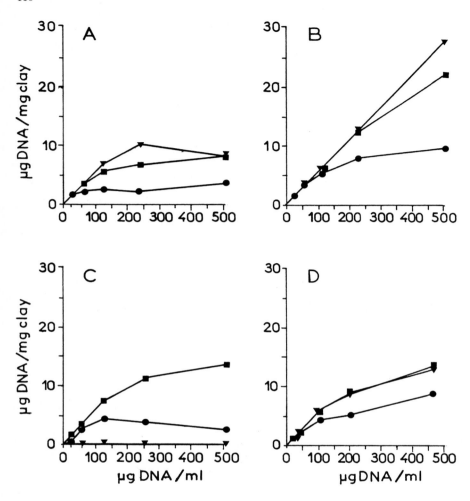

Fig. 3. Isotherms of DNA adsorption to Na-bentonite (▼), kaolinite (■) and Ca-bentonite (●) in the presence of 5 mM $MgCl_2$ (A), 200 mM $MgCl_2$ (B), 0.1 M NaCl (C) and 5 M NaCl (D)

The sand/clay system was loaded with DNA and was then treated with DNaseI. Table 2 (I) shows that the adsorbed DNA resisted degradation by DNaseI. For comparison, 6 μg of DNA in 0.2 ml solution (this volume corresponds to the interstitial volume of the sand/clay bed) were completely converted to acid soluble material within 1 h by 100 μg DNaseI ml^{-1}. Extension of the incubation time with DNaseI in the microcosm to 24 h did not result in further degradation indicating that the DNaseI provided its nucleolytic activity essentially during the first hour of incubation (Table 2). Presumably part of the DNA molecules was not reached by the enzyme or the enzyme molecules were rapidly inactivated during or after their binding to the mineral surfaces. The high sorption capacity of clay for DNA shown before (see Fig. 3A) was also noticed with the microcosm. About 60% of the applied DNA adsorbed to NaB which provided only 0.6% (w/w) of the minerals present in the sand bed (Table 2 III). The fractions of DNA adsorbed to sand and clay were both protected from degradation by DNaseI. However, there was a somewhat higher resistance against degradation of sand-adsorbed than of clay-adsorbed DNA (Table 2, compare II and III).

Table 2. Degradation of mineral adsorbed DNA by DNaseI in the sand/clay microcosm

	Mineral fraction	µg DNA adsorbed	DNaseI treatment µg ml^{-1}	h	DNA degraded (%)
I	Total (sand + clay)	6.7	100	1	36
			100	24	46
			500	1	70
			500	24	66
II	Sand (0.7 g)	2.6	100	1	9
			100	24	24
			500	1	39
			500	24	34
III	Na-bentonite (0.004 g)	4.0	100	1	52
			100	24	56
			500	1	86
			500	24	79

Conclusions

The interaction of negatively charged DNA molecules with negatively charged clay surfaces was apparently mediated by the reduction of repulsive forces by cations. This conclusion is drawn from the fact that the amount of DNA adsorbed to NaB and CaB increased with increasing NaCl concentrations (Fig. 2A). The highly efficient binding in the presence of even low concentrations of Mg^{2+} ions (Fig. 2B), previously also observed with sand, montmorillonite and material from a natural groundwater aquifer (Lorenz and Wackernagel 1987; Romanowski et al. 1991; Greaves and Wilson 1969; Romanowski et al. in preparation), may be explained as a divalent cation bridging effect (for theory see Hesselink 1983). The high binding capacities (Fig. 2 and 3) over a wide pH range (Table 1) would lead one to expect an important contribution of clay minerals as locations in soil or sediment where DNA may accumulate after release from cells. In the model soil microcosm consisting of sand minerals and sodium-bentonite, the major part of DNA (60%) bound to the low amounts (0.6%) of clay particles (Table 2).

Binding to sand and clay in the microcosm rendered DNA resistant against DNaseI (Table 2). A similar protection against enzymatic degradation was observed when circular and linear DNA molecules adsorbed to material of a groundwater aquifer containing 0.5% clay minerals (Romanowski et al. in preparation). These results give reason to the assumption that the persistence of plasmid DNA in several soils for at least 10 days (Romanowski et al. submitted) may be the result of the protective effect of adsorption to minerals including quartz, feldspar, heavy minerals and clay. Adsorbed bacterial DNA constitutes an extracellular pool of genetic information which may be available for transformation of competent bacterial cells.

References

Carter DM, Radding CM (1971) The role of exonuclease and β–protein of phage lambda in genetic recombination. II. Substrate specificity and the mode of action of lambda exonuclease. J Biol Chem 246:2502-2510

Greaves MP, Wilson MJ (1969) The adsorption of nucleic acids by montmorillonite. Soil Biol Biochem 1:317-323

Hesselink FT (1983) Adsorption of polyelectrolytes from dilute solution. In Parfitt GD, Rochester CH (ed), Adsorption from solution at the solid/liquid interface. Academic Press, London, pp 377-412

Lorenz MG, Wackernagel W (1987) Adsorption of DNA to sand and variable degradation rates of adsorbed DNA. Appl Environ Microbiol 53:2948-2952

Lorenz MG, Wackernagel W (1988) Impact of mineral surfaces on gene transfer by transformation in natural bacterial environments. In Klingmüller W (ed), Risk Assessment for Deliberate Releases. Springer-Verlag, Berlin Heidelberg, pp 110-119

Romanowski G, Lorenz MG, Wackernagel W (1991) Adsorption of plasmid DNA to mineral surfaces and protection against DNaseI. Appl Environ Microbiol 57:1057-1061

Stotzky G (1986) Influence of soil mineral colloids on metabolic processes, growth, adhesion, and ecology of microbes and viruses. In Soil Science Society of America (ed), Interactions of soil minerals with natural organics and microbes. SSSA Spec. Pub., Madison, pp 305-428

Characterization of Natural Transformation of the Soil Bacteria *Pseudomonas stutzeri* and *Acinetobacter calcoaceticus* by Chromosomal and Plasmid DNA

S. Bruns, K. Reipschläger, M. G. Lorenz and W. Wackernagel
Genetik, Fachbereich Biologie
Universität Oldenburg
Postfach 2503
W-2900, Oldenburg
Germany

Introduction

Recent microcosm studies strongly point at a potential for genetic transformation in the environment (for survey see Lorenz 1992). Dissemination of genes by free DNA among a natural microbial population requires several steps including (i) the release of functional DNA from donor cells, (ii) the persistence of extracellular DNA, (iii) the development of competence in the milieu of the natural environment, (iv) the uptake of free DNA into a competent recipient cell, (v) the propagation of the internalized DNA and the eventual expression of a newly gained trait (Wackernagel et al. 1992).

For the identification of factors affecting transformation in bacterial habitats it is desirable to use well characterized strains capable of developing natural competence. The gram-negative soil bacteria *Acinetobacter calcoaceticus* and *Pseudomonas stutzeri* are naturally transformable (Juni and Janik 1969; Carlson et al. 1983). They have also been shown to be transformed in soil extract and in groundwater (Lorenz and Wackernagel 1991; Lorenz et al. 1992). Here we describe some characteristics of the species regarding the transformation by chromosomal and plasmid DNA.

Materials and Methods

Strains and DNA isolation procedures.
For transformation studies *P. stutzeri* JM302 (*his-1*; Carlson et al. 1983) and *A. calcoaceticus* BD413 (*trpE27*; Juni and Janik 1969) were used. Transforming chromosomal DNA was isolated from the prototrophic strains *P. stutzeri* JM375 and *A. calcoaceticus* BD4 according to Marmur (1961). Plasmid DNA was prepared as described by Birnboim and Doly (1979).

Media and competence.
Cells were grown in LB with antibiotics (*A. calcoaceticus,* 20 µg chloramphenicol ml^{-1}; *P. stutzeri* 250 µg streptomycin ml^{-1}) for the isolation of plasmid DNA. Strains were grown in LB to competence as described by Juni and Janik (1969) and Lorenz and Wackernagel (1990). Prototrophic transformants were scored on minimal media (*A. calcoaceticus,* Cruze et al. 1979; *P. stutzeri,* Carlson et al. 1983) and plasmid transformants on LB supplemented with appropriate antibiotics (*P. stutzeri:* streptomycin 250 µg ml^{-1}; tetracycline 15 µg ml^{-1}; *A. calcoaceticus*: kanamycin 25 µg ml^{-1}, gentamycin 30 µg ml^{-1}, tetracycline 15 µg ml^{-1}, chloramphenicol and streptomycin 20 µg ml^{-1}).

Transformation.
Competent cultures (usually 0.2 to 0.4 ml) were incubated with chromosomal or plasmid DNA for 60 min at 30°C (*A. calcoaceticus*) or 90 min at 37°C (*P. stutzeri*). Thereafter DNaseI (100 µg ml^{-1} final concentration) was added and cells were incubated another 10 min before plating on selective media. In plasmid transformations, a period of expression in LB (usually 1.5 h) preceded plating. The procedures are described in detail elsewhere (Lorenz and Wackernagel 1990; Lorenz et al. 1992).

The modified plate assay (Juni and Janik 1969) was used for transformation of *P. stutzeri* by plasmids (Lorenz and Wackernagel

1991). Briefly, non-competent stationary phase JM302 cells (50 μl; $6 \cdot 10^9$ ml^{-1}) were mixed with 0.05 to 0.5 μg plasmid DNA and spotted on LB agar plates. After 2 d at 37°C the growth area was cut out by using a sterile spatula. The cells were resuspended in minimal medium by vigorous vortexing. After DNaseI treatment for 10 min, cells were plated on selective media containing antibiotics. Transformant colonies were counted after 2 d at 37°C.

Adsorption of DNA to cells was determined in the following manner (Lorenz and Wackernagel 1990; Lorenz et al. 1992). Competent cultures were incubated with DNA for periods as indicated in Table 1. Then a portion (0.2 ml) was layered on top of an ice cold isoosmotic Percoll gradient and centrifuged at 200 g for 20 min.

Table 1. Determination of the rate-limiting step of transformation

Strain	DNaseI treatment[b]	Centrifugation[c]	Transformation frequency[a]			
			0.5 min	5 min	15 min	30 min
P. stutzeri	+	-	<0.03			1.5
	-	+	<0.10			1.3
A. calco aceticus[d]						
+ Ca^{2+}	+	-		0.29	2.90	
	-	+		4.10	6.30	
- Ca^{2+}	+	-		0.06	0.21	
	-	+		4.60	7.60	

a 10^{-7} His$^+$ (P.s.) or 10^{-4} Trp$^+$ (A.c.) per recipient cell
b 10 μg DNaseI ml^{-1} final conc.; 10 min at 30° (A.c.) or 37°C (P.s.)
c in an isoosmotic Percoll gradient for 20 min at 200 x g (4°C); thereafter cells were collected and plated on selective medium
d incubation in 25 mM Tris-HCl buffer (pH 7.5) +/- 0.5 mM CaCl$_2$.

The band of cells in the gradient was removed by a sterile Pasteur pipette. Samples were then plated on selective minimal media.

Results and Discussion

Rate limiting step of transformation.
Kinetic studies showed that transformation of *A. calcoaceticus* and *P. stutzeri* was complete after 45 min and 120 min, respectively (not shown). To see whether binding of DNA to cells or DNA uptake into the DNaseI-resistant state was the rate limiting step of transformation, the following experiments were performed. Competent cells were incubated with homologous transforming DNA for various time periods. They were then either treated with DNaseI to destroy any DNA not yet taken up or centrifuged in a Percoll gradient to remove DNA not associated with cells at the moment of sampling. Afterwards cells were plated on selective media and transformant colonies scored. The results are shown in Table 1. Transformation frequencies of DNaseI-treated and centrifuged *P. stutzeri* cells were similarly low after 30 min of incubation with DNA (Table 1, first and second entry) indicating that binding of DNA to the cell surface was rate limiting during transformation and uptake of DNA into the DNaseI-resistant state was rapid. Possibly an interaction between cells and DNA molecules was hindered by the high motility of *P. stutzeri*. Transformation of *A. calcoaceticus* depends on divalent cations (Lorenz et al. 1992). In buffer plus Ca^{2+}, DNaseI-treated cells contained a low number of transformants after 5 min which increased 10fold within the next 10 min (Table 1, third entry). Transformation frequencies of the same cells, but centrifuged in a Percoll gradient to remove non-adsorbed DNA were already high after 5 min of incubation with DNA and did not substantially increase during the next 10 min (Table 1, fourth entry). It is concluded that, in *A. calcoaceticus*, the uptake process is the rate limiting step of transformation and not the binding of DNA to cells. When Ca^{2+} was omitted, a very low transformation frequency was obtained after 5 min which did not greatly increase within the next 10 min (Table 1, fifth entry). In contrast, centrifuged cells showed a

high frequency of transformation already after 5 min similar to cells in Ca^{2+}-containing buffer (Table 1, compare sixth and fourth entry). This is due to the uptake of cell-adsorbed DNA on the agar medium which contains Mg^{2+}. Thus the uptake process and not the binding of DNA to cells was dependent on Ca^{2+}. Mn^{2+} or Mg^{2+} can replace Ca^{2+} (Lorenz et al. 1992).

Competition of heterologous with homologous DNA.
The two bacteria were compared with respect to their ability to discriminate between DNA of their own and foreign species in series of DNA competition experiments. Table 2 shows that a 10fold excess of *E. coli* DNA reduced the transformation frequency of *P. stutzeri* only by 28 %. This suggests a selective rather than a nonselective uptake of DNA which is, however, not as strict as in e.g. *Neisseria* (Dougherty et al. 1979).

In *A. calcoaceticus,* transformation levels dropped to approximately 20 % and 5 % at 4fold respectively 20 fold excess of *E. coli* DNA (Table 2). This finding indicates that, like in pneumococci and *B. subtilis*, the binding/uptake process of *A. calcoaceticus* does not discriminate between homologous and heterologous DNA.

Table 2. Influence of heterologous DNA on transformation

Strain	Source of competing DNA	Relative transformation (%) at x-fold mass excess of competing DNA x =		
		4	10	20
P. stutzeri	*E. coli* AB1157		72.0	
A. calco-	*E. coli* AB1157	19.0		5.6
aceticus	Phage P22	18.0		6.5
	Salmon testes	19.5		4.2

DNA restriction during transformation.

In the environment competent cells may take up DNA released from the same or from other species. To study the effect of DNA restriction, we used the broad host range IncQ/P4 plasmid RSF1010 and its derivative pKT210 (Bagdasarian et al. 1981). The plasmid DNA was prepared either from *E. coli* or from the species that was transformed. *P. stutzeri* was transformed by RSF1010 DNA (Table 3). When plasmid DNA was prepared from *E. coli*, the efficieny of transformation was 7fold lower than when the DNA was isolated from *P. stutzeri* (compare entry 1 and 2). This finding is in contrast to previous reports which stated that plasmid transformation was obtained only with plasmid DNA having an insert of chromosomal *P. stutzeri* DNA and when the plasmid was prepared from *P. stutzeri* instead from *E. coli* (Carlson et al. 1984). Recently transformation with plasmid DNA without chromosomal insert was also observed by M. Mieschendahl (personal communication).

Table 3. Influence of the source of the plasmid DNA and of the expression time on transformation

Strain	Source of plasmid DNA	Plasmid (marker)	Transformation efficiency (resistant cells μg DNA^{-1} 10^9 cells^{-1})
P. stutzeri	*P. stutzeri*	RSF1010 (Sm^R)	119
	E. coli	RSF1010 (Sm^R)	18
A. calco-aceticus	*A. calco-aceticus*	RSF1010 (Sm^R)	0.15 . 10^4 (1.5 h)[a]
			5.60 . 10^4 (17 h)[a]
		pKT210 (Cm^R)	8.10 . 10^4 (1.5 h)[a]
			7.40 . 10^4 (17 h)[a]
	E. coli	pKT210 (Cm^R)	3.80 . 10^4 (1.5 h)[a]
			4.10 . 10^4 (17 h)[a]

[a] denotes expression time

An influence of the source of the plasmid DNA was even less evident in *A. calcoaceticus* transformations. The efficiency of transformation by pKT210 DNA from *E. coli* was only twofold lower than by pKT210 DNA from *A. calcoaceticus* (Table 3, compare entry 5 with 7 and 6 with 8). The results support the conclusion that, in *P. stutzeri* and *A. calcoaceticus*, DNA restriction plays only a minor role, if any. A similar mild effect of restriction was observed in *B. subtilis* transformation (Ganesan 1982).

Marker expression.

Initial studies on transformation of *A. calcoaceticus* with several plasmids revealed substantial lower frequencies of RSF1010 than pKT210 transformants. The plasmids are identical except that pKT210 contains an insert in the RSF1010 moiety. This 3.5 kb *PstI* fragment of pSa replaces the small RSF1010 *PstI* fragment and confers chloramphenicol resistance (Bagdasarian et al. 1981). The difference in the transformation efficiency of RSF1010 and pKT210 was abolished when the time of expression of markers was extended (Table 3). A 37fold higher level of transformation was obtained after 17 h of expression of the streptomycin resistance marker than after 1.5 h of expression (Table 3; compare entries 3 and 4). Further uptake of DNA during these periods was prevented by DNaseI treatment of cells 45 min after addition of DNA. Also, the cell titre did not increase during the 17 h of incubation due to the high concentration of cells ($1.3 . 10^9$ ml^{-1}). With pKT210 the number of transformants was high already after 1.5 h of expression and did not increase during the extended period of expression (Table 3; compare entry 5 with 6). Apparently the streptomycin resistance marker was expressed more slowly than the chloramphenicol resistance marker.

Survey of plasmids transforming P. stutzeri and A. calcoaceticus.
Table 4 lists those plasmids that were included in the transformation studies. The physical presence of plasmids in transformants was verified by clone analysis. *P. stutzeri* was transformed by RSF1010 and its nonmobilizable derivative pKT261 (Bagdasarian et al. 1981) and by the new shuttle vector pSBA5 (Fig. 1). This plasmid was obtained as a ligation product of *HindIII* linearized pRK2501, which

is an RP4 derivative (Kahn et al. 1979), and the *HindIII* linearized *Staphylococcus aureus* plasmid pC194 (Iordanescu et al. 1975). pSBA5 replicates also in *E. coli* and *B. subtilis*.

A. calcoaceticus was transformed by RSF1010 and the derivative thereof, pKT210 (Table 4). In addition, transformants were obtained with pSUP102-Tn5B20 which constitutes of a pACYC184 replicon (Simon et al. 1989). Because pACYC184 is a narrow host range plasmid typically replicating in enteric bacteria, pSUP102-Tn5B20 is recommended as a suicide vector for Tn5 insertion mutagenesis in gram-negative bacteria (Simon et al. 1989). The ability of pSUP102-Tn5B20 to replicate in *A. calcoaceticus* excludes this system as a mutagen in this organism. Another interesting observation was that no transformants were obtained with pRK2501 and pSBA5 (Table 4). A lack of expression of the RP4 tetracycline resistance determinant can be excluded as the reason because *A. calcoaceticus* cells harboring RP4 were Tc[R]. Even with an extension of the expression period (17 h) transformants were not obtained.

Table 4. Transforming activity of several plasmids

Plasmid tested	Size (kb)	Selected marker	Transformants μg plasmid^{-1} 10^9 cells^{-1}	
			P. stutzeri[a]	*A. calcoaceticus*[b]
RSF1010	8.9	Sm[R]	$1.2 . 10^2$	$5.6 . 10^4$
pKT261	8.6	Sm[R]	$1.1 . 10^2$	N.D.[c]
pKT210	11.8	Cm[R]	N.D.[c]	$7.4 . 10^4$
pRK2501	11.1	Tc[R]	N.D.[c]	$<7.3 . 10^1$
pSBA5	14.0	Tc[R]	$3.6 . 10^1$	$<7.3 . 10^1$
pSUP102-Tn5B20	16.5	Km[R],Gm[R]	N.D.[c]	$3.8 . 10^3$

[a] plate transformation assay
[b] transformation of competent cells in liquid
[c] N.D.= not determined

Also, the rather large size of pRK2501 and pSBA5 is probably not the reason for the failure because *A. calcoaceticus* was transformed by other plasmids of similar size (pKT210, pSUP102-Tn5B20; see Table 4). Most probably pRK2501 and pSBA5 do not replicate in *A. calcoaceticus*.

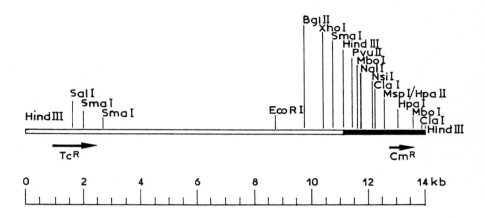

Fig. 1. Restriction map of pSBA5. The *P. stutzeri-B. subtilis* shuttle vector consists of *HindIII* ligated pRK2501 (open line) and pC194 (closed line)

Conclusions

The examination of the natural transformation of two Gram-negative soil bacteria disclosed that considerable differences exist between them in the DNA-cell interactions and in the uptake and propagation of DNA. In *P. stutzeri*, transformation appears to be selective for homologous DNA at the point of DNA entry which is otherwise typical for *Haemophilus* and gonococci (for review see Smith et al. 1981). The selective uptake of homologous DNA in this soil bacterium may limit a flow of genes by free DNA within *P. stutzeri* and closely related species as was observed previously (Carlson et al. 1983; Stewart and Sinigalliano 1991). The low transformation efficiency by three plasmids (see Table 4) can also be explained by this selective type of DNA uptake in *P. stutzeri, A. calcoaceticus* seems to follow a different strategy. The transformation of this ubiquitous bacterium is reminiscent of the Gram-positive type of DNA uptake (see Smith et al. 1981) which does not discriminate between homologous and heterologous DNA. Hence *A. calcoaceticus* may take up any DNA, including plasmids, in the environment. This may eventually lead to the acquisition of new traits. The DNA may be also used as high Mr source of C, N, and P.

With respect to the release of genetically engineered microorganisms to the environment two other findings are of interest. In *A. calcoaceticus*, the streptomycin resistance marker of RSF1010 was expressed much slower than the chloramphenicol resistance marker of pKT210 (Table 3). This may be a warning that the flow of genetic information in the environment may be underestimated if only the phenotypic detection of traits is applied for monitoring. A marker may not be expressed at all in one strain and may reappear later as a genetic trait upon transfer to another. Such an "escape" of genes has recently been reported (Top et al. 1990).

The ability of pSUP102-Tn5B20 to transform *A. calcoaceticus* (Table 4) implies that the frequently used cloning vectors of the pACYC-group (Chang and Cohen 1978) may be naturally taken up and propagated in this strain. This is of relevance to considerations regarding the safety of these nonmobilizable plasmids because *A. calcoaceticus* has been shown to be transformed by plasmid DNA in environmental samples (Lorenz et al. 1992). The data and conclusions presented here underline the notion that for studies of the environmental bacterial gene flux it is necessary to examine more than a single "model" species.

References

Bagdasarian M, Lurz R, R̦ckert B, Franklin FCH, Bagdasarian MM, Frey J, Timmis KN (1981) Specific-purpose plasmid cloning vectors. II. Broad host range, high copy number, RSF1010-derived vectors, and a host-vector system for gene cloning in *Pseudomonas*. Gene 16:237-247

Birnboim HC, Doly J (1979) A rapid alkaline extraction procedure for screening recombinant plasmid DNA. Nucleic Acids Res 7:1513-1523

Carlson CA, Pierson LS, Rosen JJ, Ingraham JL (1983) *Pseudomonas stutzeri* and related species undergo natural transformation. J Bacteriol 153:93-99

Carlson CA, Steenbergen SM, Ingraham JL (1984) Natural transformation of *Pseudomonas stutzeri* by plasmids that contain cloned fragments of chromosomal deoxyribonucleic acid. Arch Microbiol 140:134-138

Chang ACY, Cohen SN (1978) Construction and characterization of amplifiable multicopy DNA cloning vehicles derived from P15A cryptic miniplasmid. J Bacteriol 134:1141-1156

Cruze JA, Singer JT, Finnerty WR (1979) Conditions for quantitative transformation in *Acinetobacter calcoaceticus*. Curr Microbiol 3:129-132

Dougherty TJ, Asmus A, Tomasz A (1979) Specificity of DNA uptake in genetic transformation of gonococci. Biochem Biophys Res Commun 86:97-104

Ganesan AT (1982) Uptake, restriction, modification and recombination of DNA molecules during transformation in *B. subtilis*. In Ganesan AT, Chang S, Hoch JA (eds), Molecular cloning and gene regulation in bacilli. Academic Press, New York, pp 261-268

Iordanescu S (1975) Recombinant plasmid obtained from two different, compatible staphylococcal plasmids. J Bacteriol 124:597-601

Juni E, Janik A (1969) Transformation of *Acinetobacter calcoaceticus (Bacterium anitratum)*. J Bacteriol 98:281-288

Kahn M, Kolter R, Thomas C, Figurski D, Meyer R, Remaut E, Helinski DR (1979) Plasmid cloning vehicles derived from plasmids ColE1, R6K and RK2. Meth Enzymol 68:268-280

Lorenz MG (1992) Gene transfer via transformation in soil/sediment environments. Microb Releases (this issue)

Lorenz MG, Reipschläger K, Wackernagel W (1992) Plasmid transformation of naturally competent *Acinetobacter calcoaceticus* in non-sterile soil extract and groundwater. Arch Microbiol 157:

Lorenz MG, Wackernagel W (1990) Natural genetic transformation of *Pseudomonas stutzeri* by sand-adsorbed DNA. Arch Microbiol 154:380-385

Lorenz MG, Wackernagel W (1991) High frequency of natural genetic transformation of *Pseudomonas stutzeri* in soil extract supplemented with a carbon/energy and phosphorus source. Appl Environ Microbiol 57:1246-1251

Marmur J (1961) A procedure for isolation of deoxyribonucleic acid from microorganisms. J Mol Biol 3:208-218

Simon R, Quandt J, Klipp W (1989) New derivatives of transposon Tn5 suitable for mobilization of replicons, generation of operon fusions and induction of genes in Gram-negative bacteria. Gene 80:161-169

Smith HO, Danner DB, Deich RA (1981) Genetic transformation. Ann Rev Biochem 50:41-68

Stewart GJ, Sinigalliano CD (1991) Exchange of chromosomal markers by natural transformation between the soil isolate, *Pseudomonas stutzeri* JM300, and the marine isolate, *Pseudomonas stutzeri* strain ZoBell. Antonie van Leeuwenhoek 59:19-25

Top E, Mergeay M, Springael D, Verstraete W (1990) Gene escape model: transfer of heavy metal resistance genes from *Escherichia coli* to *Alcaligenes eutrophus* on agar plates and in soil samples. Appl Environ Microbiol 56:2471-2479

Wackernagel W, Romanowski G, Lorenz MG (1992) Studies on gene flux by free bacterial DNA in soil, sediment and groundwater aquifer. In Stewart-Tull DES, Sussman M (eds), The release of genetically engineered microorganisms. Plenum Publishing Corporation, New York (in press)

The Importance of Retromobilization to Gene Dissemination and the Effect of Heavy Metal Pollution on Retromobilization in Soil

E. Top, P. Vanrolleghem, H. De Rore, D. Van der Lelie[1], M. Mergeay[1] and W. Verstraete
Laboratory of Microbial Ecology
University of Gent
Coupure Links 653
B-9000, Gent
Belgium

Introduction

A phenomenon that warrants attention in the context of gene dissemination in the environment is retromobilization. Retromobilization (or retrotransfer) is the ability of some conjugative plasmids (in particular IncP1 plasmids) to mobilize genes in the opposite direction (from recipient to donor of the conjugative plasmid). This phenomenon has been observed for chromosomal markers, non-mobilisable and mobilisable vectors (Mergeay et al. 1987; Mergeay et al. 1990; Top et al. 1990). As retromobilization appears to be a way for bacteria to capture new genes from other species, an enhanced spread of introduced genetic material into a microbial community could be expected when retromobilizing plasmids are present among the indigenous microbiota.

[1] Laboratory of Genetics and Biotechnology. Vlaamse Instelling voor Technologisch Onderzoek (VITO), Boeretang 200, B-2400 Mol, Belgium

This paper aims at giving a short discussion of two methods used to characterize the retromobilization mechanism. Using two different approaches, we investigated the hypothesis, arisen during earlier research, that retromobilization of genes occurs in one step, through the pilus formed by the Tra^+ plasmid while present in its original host. In addition, the ecological significance of retromobilization was studied in soil microcosms and the effect of heavy metal pollution on retromobilization of heavy metal resistance genes was assessed.

Characterization of retromobilization by use of mathematical modelling

The kinetics of retromobilization were studied according to existing mathematical models for conjugal transfer, based upon a mass action approach: the rate of transconjugant formation is jointly proportional to the concentrations of donor and recipient cells (Freter et al. 1983; Levin et al. 1979).

Our first model ("one-step model") describes the hypothesis that retromobilization occurs in one step, and thus has the same kinetics as a direct plasmid transfer (Fig 1a). Mass action leads to the equation:

$$dN/dt = k_{12} \times R \times D \qquad (1)$$

where t is time (hours), D, R and N respectively the concentration (CFU/ml) of donors of the Tra^- plasmid, recipients harbouring the Tra^+ plasmid, and "retrotransconjugants", k_{12} the conjugal transfer rate constant (h^{-1} $(CFU/ml)^{-1}$) of the one-step retromobilization of the Tra^- plasmid into the recipient (R). Other processes occurring at the same time (like decrease of D and R) are not discussed here, as a detailed study is beyond the scope of this short article, but can be neglected (Top et al. 1992b). The analytical solution for N can be simplified to the following equation:

$$N = k_{12} \times R \times D \times t \qquad (2)$$

showing a linear relation between N and R, or N and t.

In the second model, called the "two-step model", "donor-transconjugants" D' are formed in a first stage; in a second stage "retro-transconjugants" are formed as a result of encounters between D' and R (Fig. 1b).

This model is based upon two differential equations:

$$dD'/dt = k_1 \times R \times D \qquad (3)$$

$$dN/dt = k_2 \times R \times D' \qquad (4)$$

with k_1 representing transfer of the Tra$^+$ plasmid to D, and k_2 representing mobilization of the Tra$^-$ plasmid from D' to R. Integration of the above equations, and reduction of the obtained equations leads to:

$$N = k_1 \times k_2 \times R^2 \times D \times t^2/2 \qquad (5)$$

In this model, the relation between N and R, or N and t, is not longer linear but quadratic.

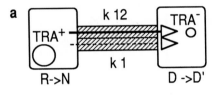

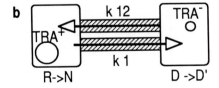

Fig. 1. Schematic representation of retromobilization of a Tra$^-$ Mob$^+$ plasmid by a Tra$^+$ plasmid according to the one-step (a) and two-step model (b). For explanation of symbols: see text.

As the dependence of N on R is different for the two models, filter matings with *E. coli* harbouring the Tra$^+$ IncP1 plasmid RP4 (strain R) and *E. coli* containing the Tra$^-$ Mob$^+$ vector pSUP202 (strain D) were performed with several initial recipient concentrations R, and the number of retrotransconjugants N was determined. In Fig. 2 the number of retrotransconjugants is plotted against the recipient number on a logarithmic scale. The slope of the curve of the two-step model is twice the slope of the one-step model. Fig. 2 clearly demonstrates that our data fit perfectly with the one-step model, and not with the two-step model.

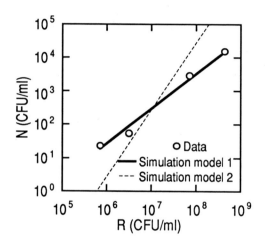

Fig. 2. Number of retrotransconjugants N obtained after 1 hour mating, plotted against the number of recipients.

Characterization of retromobilization by use of a conjugation deficient mutant

E. coli ED8739 (Murray et al. 1977) and its mutant defective as recipient in conjugations with an IncP1-type donor, isolated by Van der Lelie et al. (1992), were used in mobilization and retromobilization experiments with RP4 as Tra$^+$ plasmid, pSUP202 as Tra$^-$ Mob$^+$ vector, and CM120 (Bachmann 1972) as mating partner. As a consequence of cell wall alterations, this mutant appears to receive plasmids at a lower frequency during conjugation with IncP1 plasmids, and provides an appropriate tool to investigate the direction of pilus formation during retromobilization. Briefly, the direct mobilization of pSUP202 and the transfer of RP4 towards the mutant were 50, respectively 90 times lower than to ED8739. Matings with ED8739 or its mutant as donor of both plasmids revealed that the mutant is not affected in its donor abilities. The frequency of retromobilization of pSUP202 towards the mutant was equal to the retromobilization frequency to ED8739 (4 x 10^{-6} /recipient), indicating that the pilus is only formed from ED8739(RP4), or its mutant, towards CM120(pSUP202), and not in the opposite direction (Top et al. 1992a).

The impact of heavy metal pollution on the retro-mobilization of heavy metal resistance genes in soil.

A model system has been developed that simulates the release of cloned genes in a soil environment, based on the conjugal transfer of heavy metal resistance genes (*czc* genes, encoding resistance to Co, Zn, Cd). Direct, triparental and retromobilization of these genes, cloned in a nonconjugative (Tra$^-$ Mob$^+$) vector (IncP1), and in a nonmobilisable vector (pBR325) have been studied between *Escherichia coli* and *Alcaligenes eutrophus* in plate matings and partially in soil microcosms (Top et al, 1990). In order to investigate if heavy metal pollution has an influence on the dissemination of genes conferring resistance to this pollution, direct mobilization of the *czc* genes from *E. coli* to *A. eutrophus* in heavy metal polluted soil was compared with mobilization in non-polluted soil. In sterile polluted soil a significant increase in transconjugant numbers (up to 75% of the recipients) was observed. This proliferation of resistant transconjugants did not occur in non-polluted soil (De Rore et al. 1992). Retromobilization of the *czc* genes, cloned in a Tra⁻ Mob⁺ IncQ vector (pKT210) by means of RP4 was also studied in polluted and non-polluted soil microcosms. Again, a positive effect of heavy metal pollution on the proliferation of transconjugants was clearly demonstrated. The same microcosm studies were performed with *P. putida* as donor, simulating the deliberate release of GEMs in the environment and similar effects were noticed (Fig. 3). Retromobilization was also observed in non-sterile soil.

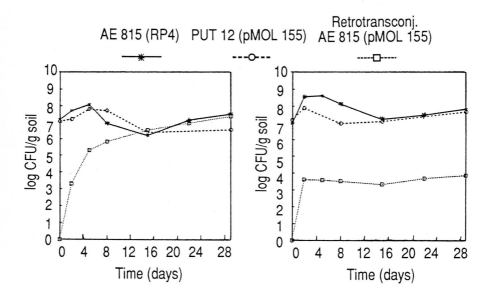

Fig. 3. Retromobilization of the IncQ plasmid pMOL155 (=pKT210::*czc*) by means of RP4 between *P. putida* and *A. eutrophus* in sterile polluted (a) and non-polluted (b) soil. AE 815 = *A. eutrophus*, PUT12 = *P. putida*

Conclusions

The results of both methods demonstrate that retromobilization of Tra⁻ Mob⁺ plasmids appears to occur in one step, through one pilus, and thus clearly differs from the process of triparental mobilization during which two encounters are needed. The ecological significance of retromobilization was investigated in soil microcosms. Retromobilization occurs in both sterile and non-sterile soil and the dissemination of the introduced *czc* genes is enhanced by the presence of heavy metals. This indicates that a recipient strain capturing a plasmid that provides the strain with a selective advantage, can become dominant under an appropriate selection pressure.

Acknowledgements

E. Top is Research Assistant of the Belgian National Fund for Scientific Research

References

Bachmann BJ (1972) Pedigrees of some mutant strains of *Escherichia coli* K-12. Bacteriol Rev 36:525-557

De Rore H, Top E, Mergeay M, Verstraete W (1992) The effect of heavy metal pollution on the dissemination of heavy metal resistance genes in soil microcosms Submitted to Appl Environ Microbiol

Freter R, Freter RR, Brickner H (1983) Experimental and mathematical models of *Escherichia coli* plasmid transfer *in vivo* and *in vitro*. Infect Immun 39:60-84.

Levin BR, Stewart FM, Rice VA (1979) The kinetics of conjugative plasmid transmission: fit of a simple mass action model. Plasmid 2:247-260.

Mergeay M, Lejeune P, Sadouk A, Gerits J, Fabry L (1987) Shuttle transfer (or retrotransfer) of chromosomal markers mediated by plasmid pULB113. Mol Gen Genet 209:61-70.

Mergeay M, Springael D, Top E (1990) Gene transfer in polluted soils. In J.C. Fry and M.C. Day (eds), Bacterial genetics in natural environments. Chapman & Hall, London, New York, p. 152-171.

Murray NE, Brammar WJ, Murray K (1977) Lambdoid phages that simplify the recovery of *in vitro* recombinants. Mol Gen Genet 150: 53-61

Top E, Mergeay M, Springael D, Verstraete W (1990) Gene escape model: transfer of heavy metal resistance genes from *Escherichia coli* to *Alcaligenes eutrophus* on agar plates and in soil samples. Appl Environ Microbiol 56:2471-2479.

Top E, van der Lelie D, Verstraete W, Mergeay M (1992a) Characterization of the retromobilization phenomenon by use of a mutant defective in conjugation with a P1-type donor (ConP⁻) In preparation.

Top E, Vanrolleghem P, Mergeay M, Verstraete W (1992b) Characterization of the retrotransfer phenomenon using mechanistic mathematical modeling. Submitted to J Bacteriol

Van der Lelie D, Top E, Nuyts G, Mergeay M (1992) Isolation of a mutant defective in conjugation with a P-type donor (ConP⁻) that is lacking the major outer membrane protein OmpF. In preparation.

Interactions between Actinophage and their Streptomycete Hosts in Soil and the Fate of Phage Borne Genes

P. Marsh and E.M.H. Wellington
Department of Biological Sciences,
University of Warwick,
Coventry, CV4 7AL.
UK

Introduction

Streptomycetes are an integral part of the soil community and grow as branching mycelia within soil aggregates. In the presence of exogenous nutrients, soil is rapidly colonized by growing mycelia, and the subsequent reduction of nutrients results in spore formation. The spore stage acts as a desiccation and phage-resistant dispersal mechanism. The combination of mycelial growth and dispersal of spores through soil are significant in the spread of genes in a particular population (Wellington et al. 1990). The major roles of streptomycetes in soil include the later stages of decomposition of recalcitrant polymers such as cellulose, lignin, chitin and starch. There is also evidence of a role in pathogen suppression in the rhizosphere via the secretion of antimicrobial compounds. Studying the population ecology of streptomycetes, and their interactions with other bacteria and fungi in the natural environment will give insights into the survival and spread of introduced genes. Such studies may enable predictions to be made about the fate of released genetically engineered microorganisms (GEMs).

Phage replication and abundance depend upon the presence of susceptible hosts. Herron and Wellington (1990) demonstrated that an increase in free actinophage titre in soil corresponded to the germination and mycelial growth of streptomycete hosts. The following experiments were designed to study the ecology of the temperate actinophage φC31 KC301 and its streptomycete host. This phage-host system may be regarded as a "predator-prey" model, which may be self-regulatory so as to ensure survival of the host and phage. This implies a coexisting steady state which is density dependant (Alexander 1981).

Spread of genes from φC31 KC301 into indigenous hosts in nonsterile unamended soil

The temperate actinophage φC31 is the most studied streptomycete phage, which infects about half of the 137 strains tested and may lysogenize 30-40% of a particular host population (Chater 1986). The derivative φC31 KC301 contains the thiostrepton resistance gene *tsr* which is expressed when susceptible hosts are lysogenized. The purpose of this experiment was to determine whether φC31 KC301 would replicate and survive in the presence of indigenous hosts and other natural populations, and whether by demonstrating lysogeny using the *tsr* encoded phenotype, the phage could survive in the prophage state in indigenous hosts.

Methods
Batch microcosms of air-dried Warwick soil (as described in Herron and Wellington, 1990) were inoculated with a phage suspension in sterile distilled water to give a free phage density of 4.0×10^5 pfu/g and between 13 and 14 % total moisture (about 40 % of moisture holding capacity). The microcosms were incubated for 15 days at 22°C, with free phage and host densities being monitored over this period. Total streptomycete counts (spores plus mycelia) were enumerated by extracting from 1g of soil in 9ml of 1/4 strength Ringer's solution, and plating the extract on reduced arginine starch salts agar (RASS) (Herron and Wellington 1990) containing rifampicin (Sigma) 10 µg/ml; nystatin (B.D.H.) 50 µg/ml; and cycloheximide (Sigma) 50 µg/ml, and incubating for 4-5 days at 30°C. Lysogens were selected for on RASS containing thiostrepton (Sigma) (50 µg/ml) in addition to the other additives. Spores were

selectively isolated using the method of Herron and Wellington (1990), the extracts being treated in the same way as for the total host counts. Free actinophages were extracted and enumerated using method C of Lanning and Williams (1982), the lawn being a pure culture of *Streptomyces lividans* TK24.

Results

Lysogenic spores of indigenous streptomycetes were isolated from the microcosms from day 2 onwards (Fig. 1). This proved that the soil under study contained host populations capable of being infected by ϕC31 KC301, and therefore able to support replication of this actinophage. Stability of these lysogens was tested by sub-culturing them on a selective medium (containing thiostrepton). Out of 22 isolates, 2 had lost thiostrepton resistance after 9 generations (superinfection resistance and molecular evidence of lysogeny will be investigated). A small proportion of the lysogens were therefore unstable, the phage DNA segregating from the host DNA after several rounds of replication, or they are pseudolysogens.

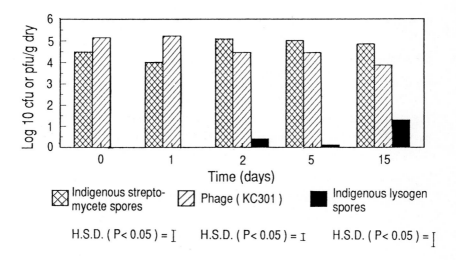

Fig. 1. Recovery of indigenous lysogen spores from non-sterile unamended soil inoculated with ϕC31 KC301

Phage and host densities at which interactions take place in sterile unamended soil

The aim of this experiment was to determine that there are minimum densities of two interacting soil populations below which no interaction can take place.

Methods

S. lividans TK24 (streptomycin resistant) was added in sterile distilled water to batch microcosms of sterile unamended soil to give a density of 1.6×10^6 CFU/g, and about 40 % of moisture holding capacity. To a range of these microcosms, ϕC31 KC301 was added at densities between 1.3×10^0 and 1.3×10^4 PFU/g. To further microcosms, ϕC31 KC301 was added at 2.0×10^4 PFU/g, with *S. lividans* TK24 added to a range of these at between 2.0×10^0 and 2.0×10^5 CFU/g. Free phage, spore and total counts were monitored as described earlier, enumeration being made on RASS containing streptomycin (Sigma) (10 µg/ml). Lysogens were selected for on RASS containing thiostrepton (50 µg/ml) and streptomycin.

Putative lysogeny in all experiments was confirmed by demonstrating spontaneous phage release from colonies picked on to overlays of *S. lividans* TK24, and by colony hybridization with radio-labelled ϕC31 KC301 DNA.

Results

The minimum phage density for phage replication (in the presence 1.6×10^6 CFU/g host) was 1.3×10^3 PFU/g, and this was also the minimum density at which lysogeny occurred (Fig. 2). Similarly, the minimum host density for phage replication and the incidence of lysogeny (in the presence of 2.0×10^4 PFU/g phage) was 2.0×10^3 CFU/g (Fig. 3). No host or phage were detected below 10^2 CFU/g or PFU/g.

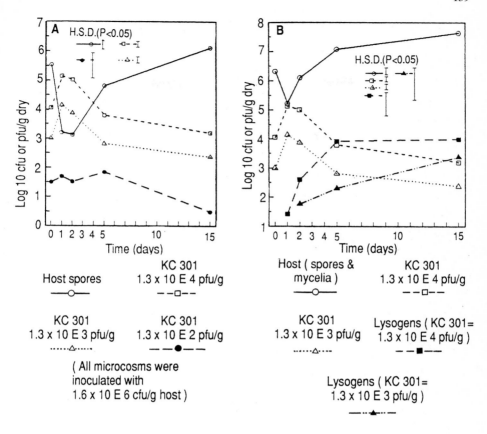

Fig. 2. Sterile unamended soil inoculated with host and varying phage densities.
A. Population changes of *S. lividans* TK24 spores and φC31 KC301
B. Population changes in total *S. lividans* TK24 and φC31 KC301 and the appearance of lysogens.

140

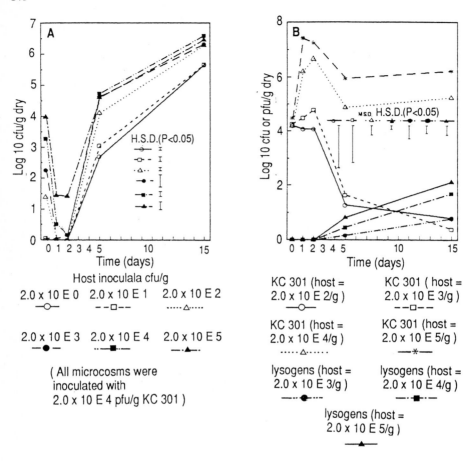

Fig. 3. Sterile unamended soil inoculated with φC31 KC301 and varying host densities.
A. Population changes of *S. lividans* TK24 spores
B. Population changes in φC31 KC301 and appearance of lysogens spores.

Discussion and Conclusions

Lysogeny of host populations indigenous to Warwick soil was possible, and this implied that replication and survival of the phage in the natural environment was also possible. The role of pseudolysogeny in φC31 KC301's life cycle may have a significant and separate role to the phenomenon of true lysogeny, in that it may facilitate a quick re-introduction mechanism into the free state for a proportion of a phage population, whilst another proportion is held in a steadier dormant state as true lysogens.

Density dependance appeared to be operating with regard to detectable interactions between ϕC31 KC301 and *S. lividans* TK24 in sterile unamended soil. That is, lysogeny of the host and replication of the phage was reliant upon the presence of sufficient densities of the host and phage (Wiggins and Alexander 1985). In every case, after the burst in phage numbers in soil which corresponded to the germination of the spores and mycelial growth, the phage numbers usually dropped by about one magnitude, corresponding with re-sporulation of the host. After this, the phage density did not alter significantly as monitored over the 10 days following the replication period. Why does the free phage density not remain at the higher level achieved during replication? Perhaps the sharp drop in free phage numbers seen with the host re-sporulation indicated a change in equilibrium between phage particles attaching to host and phage particles being released from host, *i.e.* re-sporulation itself may directly reduce the free phage population. This might be a density dependance mechanism, ensuring that the exogenous phage population did not build up to a level at which host numbers would begin to be reduced by the lytic cycle of temperate phage.

There are clearly many mechanisms which operate to ensure the survival of the phage-host system in soil, only several of which have been investigated in this work.

Acknowledgement

This project is funded by the Natural Environment Research Council.

References

Alexander M (1981) Why microbial predators and parasites do not eliminate their prey and hosts? Ann Rev Microbiol 35:113-133.

Chater KF (1986) Streptomyces phage and their applications to Streptomyces genetics, pp 119-158. In *The Bacteria, vol IX, The antibiotic producing streptomycetes.* Edited by SW Queener and LE Day. New York: The Academic Press, Inc.

Herron PR, Wellington EMH (1990) New method for the extraction of streptomycete spores from soil and application to the study of lysogeny in sterile amended and nonsterile soil. Appl Environ Microbiol 56:1406-1412.

Lanning S, Williams ST (1982) Methods for the direct isolation and enumeration of actinophages in soil. J Gen Microbiol 128:2063-2071.

Wellington EMH, Cresswell N, Saunders VA (1990) Growth and survival of streptomycete inoculants and extent of plasmid transfer in sterile and nonsterile soil. Appl Environ Microbiol 56:1413-1419.

Wiggins BA, Alexander M (1985) Minimum bacterial density for bacteriophage replication: implications for significance of bacteriophages in natural ecosystems. Appl Environ Microbiol 49:19-23.

Assessment of the Potential for Gene Transfer in the Phytosphere of Sugar Beet

M.J. Bailey, N. Kobayashi, A.K. Lilley, B.J. Powell and I.P. Thompson
Natural Environment Research Council
Institute of Virology and Environmental Microbiology,
Mansfield Road
Oxford, OX1 3SR
UK.

A comprehensive understanding of the microbial ecology of the phytosphere of sugar beet is required to make valid assessments concerning the fate of genetically modified microorganisms (GMOs) and their DNA in this habitat. Descriptions of resident populations are considered necessary as indigenous communities are potential mediators of *in situ* gene transfer events to and from introduced GMOs. In order to model impact analysis a survey, over an entire growing season, of the bacteria, yeast and filamentous fungi present on the aerial surface of field grown sugar beet (*Beta vulgaris* var. amethyst, Germians seeds, U.K), was undertaken. Bacteria were isolated on Tryptic Soy Agar (Difco) and identified to the species level by gas chromatography (Hewlett Packard 5096A) of component cellular fatty acid methyl esters using a commercially available database (MIDI-MIS, Delaware, USA). Figures 1 and 2 show the temporal and spatial distribution of the natural background bacterial populations. Members of the *Pseudomonadaceae* and *Enterobacteriaceae* were the most abundant colonizers isolated and within these groups, subtle fluctuations in the distribution of component species were recorded.

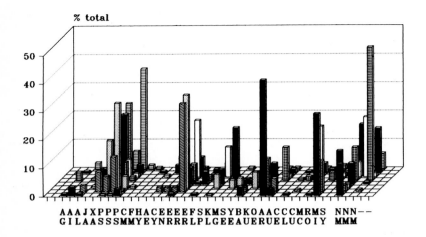

% total

Fig 1. Temporal change in the relative numbers of identified bacterial genera isolated from the phyllosphere of sugar beet plants over an entire growing season.

Key :AG-*Agrobacterium;* AI-*Acinetobacter*; AL-*Alcaligenes*; JA-*Janthinobacterium*; XA-*Xanthomonas*; PS-*Pseudomonas* other than PS-a and PS-s; PS-a *P.aureofaciens*; PS-s *P.syringae*; CM-*Comamonas;* FM-*Flavimonas*; HY-*Hydrogenophaga*; AE-*Aeromonas*; CY-*Cytophaga*; EN-*Enterobacter*; ER-c *Erwinia chrysanthemi*; ER-h *E. herbicola*; ER-r *E. rhapontici*; FL-*Flavobacterium*; SP-*Sphingobacterium*; KL-*Klebsiella*; MG-*Morganella*; SE-*Serratia*; YE-*Yersinia*; BA-*Bacillus*; KU-*Kurtobacter*; OE-*Oerskovia*; AR-*Arthrobacter*; AU-*Aureobacterium*; CE-*Cellulomonas*; CL-*Clavibacter*; MC-*Microbacterium*; RO-*Rhodococcus*; MI-*Micrococcus*; SY-*Staphylococcus*; NM-unidentified isolates.

See legend Fig. 2 for bar key of sampling times (15 to 272 days).

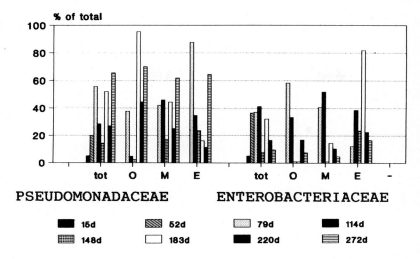

Fig 2. Spatial and temporal distribution of named genera isolated from sugar beet phyllosphere. Data presented as percentage of total bacteria isolated from each leaf type at each sampling time.
Symbols : tot- total phyllosphere; individual leaf types, O-oldest, M-mature fully expanded, E-emerging. Plants differentiated 79 days after planting. On day 148 a single species of an unidentified Coryneform (code NM-37) was abundant on the O & M sample (74% & 68% of total).

The presence of plasmid DNA in indigenous isolates, collected from mature plants prior to harvest, was determined by gel electrophoresis of DNA extracted by the method of Kado and Liu (1981) or Birnboim and Doly (1979). Of the 435 bacterial isolates examined 78 (18%) contained plasmids Fig 3 compares the PstI plasmid restriction pattern profiles and chromosomal fingerprints of a selection of isolates taken from the same leaf sample.

146

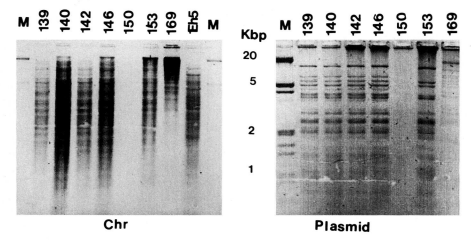

Chr   Plasmid

Fig. 3. 0.8% agarose gel in TBE (75V, 3h); PstI REN digested chromosomal and plasmid DNA isolated from *Erwinia* sp. 139, 140, 142, 146, 153 = *E. herbicola*; 150 = unidentified Gram-negative; 169 = *Serratia liquefaciens*; Eh5 = *E. herbicola*, plasmid free. These plasmid isolates shared homology with FIIA inc/rep probe.

Plasmids were extracted from both Gram-negative and Gram-positive bacteria. Fifteen percent of Pseudomonads, 38% *Enterobacteriaceae* (*Erwinia*, *Klebsiella* and *Serratia*) and up to 20% Gram-positive isolates contained plasmid DNA. Fig 3 shows chromosomally isogenic *Erwinia* sp. which vary in plasmid content, providing indirect evidence for natural plasmid movement within populations. The environmental factors that control plasmid stability and the rate of natural gene transfer are unknown. Moreover the function of many of these plasmids is cryptic. After repeated subculture on laboratory media indigenous plasmids were found to be stably maintained in their natural host. Attempts were made to characterize the 78 plasmids identified above according to homology with defined inc/rep group probes (e.g., inc/rep FIB, FIIA, 9, HI1, HI2, I1, B/O, L/M, N, P, Q, U, W, X) (Couturier et al. 1988). Only plasmids carried by members of the *Enterobacteriaceae* shared homology with the inc/rep probes. Homology was limited to the FIB and FIIA probes, 26 out of the 28 plasmid containing *Enterobacteriaceae* isolates reacted. No reactivity with any pseudomonad isolate was recorded, an indigenous pseudomonad, *Pseudomonas aureofaciens*, transformed with RSF1010 (IncQ) or RP4 (IncP) reacted strongly. This preliminary study indicates that the incompatibility groups and replication origins in plasmids collected from the phytosphere, differ from those derived from clinical sources. The biological significance of these findings is unclear. However, one can speculate that, in the absence of recognized incompatibility barriers, the stable transfer of recombinant material from introduced plasmid containing GMOs to indigenous microflora may be more likely than had previously been considered.

In terrestrial habitats (i.e., soil and plant surfaces) plasmid mediated conjugal transfer is considered to be the most common mechanism for genetic exchange between bacteria. However, few studies have investigated the distribution or abundance of promiscuous transfer plasmids indigenous to the natural terrestrial environment. Two approaches are possible, the first (endogenous) involves the extraction and characterization of plasmids by physical methods from isolated cells and the second (exogenous) directly determines the presence of transfer activity in environmental samples either by using reporter (tra$^-$ mob$^+$) plasmids or by isolating conjugative plasmids using selectable recipients (Fry and Day 1990). The natural plasmid containing isolates, described above, were assessed for their ability to mediate the mobilization of RSF1010 (tra-, mob+, Sm, Sp) in to a selectable recipient indigenous to the phylloplane of sugar beet, *P. aureofaciens* SBW25 (tetracycline resistant). Of the natural plasmid containing isolates tested, 6 (3 plasmid types) were able to mobilize RSF1010 (detection limit <1 x 10^{-8}) in to this *Pseudomonas* sp. The restriction profiles of the three tra$^+$ plasmids are shown in Fig. 3 (*Erwinia* sp. isolates, plasmidtype 139, 153 and 169). Conjugative plasmids were also isolated from bacteria indigenous to the phytosphere on the basis of their ability to express resistance to heavy metals (i.e., mercury). After mating the phytosphere extracted bacteria with recipient rif^r *P. putida* UWC1, transconjugants were selected on media containing rifampicin and mercury. Plasmid content was confirmed by extraction and gel electrophoresis. Transconjugants were obtained at frequencies between 10^{-2} and 10^{-8} for a range of isolated plasmids.

The use of alternative recipients, i.e., *Enterobacteriaceae*, was found to increase the frequency of isolation with the exogenous method and extend the diversity of samples collected.

The approaches outlined above demonstrate the importance of studying microbial population biology and plasmid ecology in the natural environment. The presence of plasmids in the indigenous natural microflora of the plant surface, which are self-tranmissible and can mediate heterologous gene transfer, has been shown. Many of the indigenous plasmids failed to react with known incompatibility groups indicating that current plasmid classification systems need to be extended.

Studies of the type described here will provide data essential for assessing the risks involved in the widespread release of GMOs into the open environment.

References

Birnboim HC, Doly J (1979) A rapid alkali extraction procedure for screening recombinant plasmid DNA. Nuc Acid Res 7:1513-1523.

Couturier FB, Berquist PL, Maas WK (1988) Identification and classification of bacterial plasmids. Microbiol Rev 52:175-195.

Fry JC, Day MJ (1990). Plasmid transfer in the epilithon. In: Eds Fry and Day "Bacterial genetics in natural environments". pp. 55-80. Chapman and Hall.

Kado C, Liu ST (1981). Rapid procedure for detection and isolation of large and small plasmids. J Bacteriol 145:1365-1373.

Section 4

HUMAN and ANIMAL GUT

Gene Transfer in Human and Animals Gut

Y. Duval-Iflah
I.N.R.A.- Centre de Recherche de Jouy
Unité Ecologie et Physiologie du Système Digestif
78 352, Jouy-en-Josas Cedex
France

Introduction

Genetic exchanges between procaryotes are mediated mainly by extrachromosomal elements, plasmids and phages. Horizontal gene exchanges can occur by conjugation, transduction or transformation. Conjugation and transduction are the must likely mechanisms of transfer between closely related species. Introduction of foreign DNA into phylogenetically remote organisms is at present performed in laboratory by means of transformation or transfection. Within the framework of the genome, movements of certain sequences and DNA rearrangement are due to transposable elements (transposons). These are responsible for plasmid and bacterial evolution. There is evidence that heterogramic genetic exchanges have occurred under natural conditions (Lambert et al. 1985; Trieu-Cuot et al. 1985; Trieu-Cuot and Courvalin 1986; Brisson-Noël et al. 1988). Genes originating in Gram-positive bacteria are readily expressed and selected in Gram-negative bacteria; the reversal polarity of exchange is performed only in laboratory with vectors. Concerning gene transfer in the gut, one have to consider gene flux between the microorganisms that colonize the gut and between procaryotes and eucaryotes. Rapid degradation of DNA in gut lumen (Maturin and Curtis 1977; Hoskins 1978), suggests that genetic exchange by direct insertion of naked DNA is unlikely between eucaryotic host cells and enteric microflora in gut lumen. Animal and human guts are colonized by a tremendous quantity of bacteria belonging to different species and families. The dominant flora is essentially composed of strict anaerobic Gram-positive and Gram-negative bacteria whereas the subdominant flora is facultative anaerobic. It is likely that genetic exchanges can occur among these bacterial populations. The experimental genetic exchanges which have been studied and demonstrated in the gut are mainly concerned with conjugation or conjugation-like systems. Experimental studies have been performed on human volunteers as well as on conventional animals, but our main understanding of the gene transfer in different ecological and genetical conditions have been obtained with the use of gnotobiotic animals.

Experimental gene transfer in the gut of human volunteers and conventional animals.

These studies get great interest since the development of genetically modified bacteria during the years 1970. Most genetically modified bacteria were originated from *Escherichia coli* of human faecal origin (*E.coli* K-12), and therefore the main concern was to know whether K-12 and its genetically modified derivatives were able to colonize efficiently and abnormally human gut and to disseminate the recombinant genetic material to resident microflora. An *E. coli* strain is designated transient when it is eliminated from digestive flora within few days or few weeks after it has been ingested ; it might become a resident when it persits for months and years (Sears et al. 1950). Survival of different *E. coli* K-12 was tested in human volunteers (Levy and Marshall 1979; Levy et al. 1980; Levy 1984; Smith 1975; Williams 1977). All these tests indicated that *E. coli* K-12 and its derivatives are not able to colonize the human digestive tract in the absence of any antibiotic selective pressure. Similar studies have been performed with conventional and axenic mice and rates treated or not with selective antibiotics (Cohen et al. 1979; Laux et al. 1982; Levy and Marshall 1979; Levy et al. 1980; Levy 1984; Myhal et al. 1982; Smith et al. 1985; Wells et al. 1978). The conclusion from these tests was that *E.coli* K-12, except the debilitated variant X1776 (Wells et al. 1978), were able to colonize germ-free mice and antibiotic treated mice and rats but not conventional animals. These results are in agreement with the assumption that the unability of any *E.coli* strain (a wild type or a K-12) to colonize the digestive tract is due to competitive effect exerted by indigenous microflora and to intraspecific antagonism exerted by other strains of *E. coli* (Duval-Iflah et al. 1980, 1981).

Concomitantly with the survival of *E. coli* in the gut, studies were achieved in order to know the transferability of different autotransferable or non autotransferable plasmids harboured by K-12 or other *E. coli* strains. The transfer of R-factors from *E.coli* and other enterobacterial strains of human and animal origin was also assessed. Non conjugative plasmids carried by K-12, like pBR322 and pBR325 , were not transferred when present alone or together with a mobilizing plasmid to other *E.coli* strains unless antibiotic pressure is exerted (Levy and Marshall 1979; Levy et al. 1980; Marshall et al. 1981; Smith 1975). In one instance ColV plasmid was transferred to *E. coli* resident strains; a triparental mating involving a conjugative plasmid from a resident strain was assumed to have occurred (Williams 1977). Some but not all conjugative plasmids harbored by K-12 can transfer to resident *E. coli* in human volunteers (Marshall et al. 1981). Several investigations have indicated spontaneous transfer of R-factors between enterobacterial strains in human gut: R-factor transfer from *E. coli* to *Shigella sonnei* (Farrar et al. 1972), from ingested *E. coli* to resident *E. coli* (Andersen 1975) and between two resident *E. coli* strains (Petroucheilou et al. 1976). R-factor transfer between strains of faecal origin was observed under antibiotic selective pressure in human gut (Anderson et al. 1973). In animal gut, R-factor transfer was also observed under natural conditions and under antibiotic selective pressure (Guinee 1965, 1968, 1970; Kazuya 1964; Gyles et al. 1978). Of most interest was the investigation undergone by Smith (1971) who compared the transferability of the same R-factor between a couple of donor (*E.coli*) and recipient (*Salmonella typhimurium*): the transfer was easy and at high frequency in chicken and in veals gut where *S. typhimurium* was able to become established, but not in piglets gut where *S. typhimurium* was not established. These results, together with the fact that antibiotics promote R-factors transfer, suggest that bacterial mating is a phenomenon that obviously happens in natural conditions in human and animal digestive tract. The reason why it is observed only on rare occasions comes from the unability of the donor and / or the transconjugant to become established durably in natural conditions.

Use of germ-free animals in the studies of gene transfer between microorganims in the gut.

Utilisation of germ-free animals allowed the study of the effect of different factors (genetical and ecological) that may control gene transfer in the gut. They were also used to determine the various ways of gene transfer which are possible in the gut.

In vivo matings in mice diassociated with recipient and donor strains.

When germ-free were associated with donor and recipient strains only, matings were observed between strains belonging to same or distinct species: *Shigella* and *Klebsiella* (Kasuya 1964), *E. coli* strains of human faecal flora (Duval et al. 1981), a strain of *Serratia* (isolated from human urinary tract) to *E. coli* of human faecal origin (Duval et al. 1980). Conjugative plasmids were able to transfer between two K-12 strains (Freter et al. 1983; Andremont et al. 1985). On the other hand, the non conjugative plasmid pBR322 carried by K-12 was not able to transfer to human faecal *E. coli* strains even though donor was at the very high level of 10^{10} per gram of feces (Levy 1984). Plasmid transfer between strains which do not belong to *Enterobacteriaceae* were also described: mating between strains of *Clostridium perfringens* (Brefort et al. 1977); transfer of plasmid pAMß1 from *Lactobacillus reuteri* to *Enterococcus faecalis* (Morelli et al. 1988), a variant of pAMß1 from *Lactococcus lactis* to *Enterococcus faecalis* (Duval-Iflah and Gruzza 1990; Gruzza et al. 1990). Conjugative transposon was transferred from *Enterococcus faecalis* to *Listeria monocytogenes* (Doucet-Populaire et al. 1991).

Parameters controlling gene transfer in the gut.

Inoculum size was shown to have an effect only in the case where the donor was not able to become established, under conventional conditions (Smith 1969) and in gnotobiotic conditions (Sansonetti et al. 1980). On the other hand, establishment of donor strains was not necessary for plasmid transfer since it was shown that donors like *Serratia* and *L. lactis*, which were drastically eliminated from the digestive tract of gnotobiotic mice, were able to transfer their conjugative plasmid to resident recipients, respectively *E. coli* and *E. faecalis* (Duval-Iflah et al. 1980, 1990; Gruzza et al. 1990). Sansonetti et al. (1980) showed that, in gnotobiotic chicken gut, both derepressed and repressed variants of the same plasmid transfer with equal and high efficiency to a recipient *E. coli* strain *con*I$^+$ and confer an ecological advantage to the transconjugants. On the other hand, the two plasmids transfer at less extent to the low efficient recipient *E. coli con*I and the transconjugants were at disadvantage. The same phenomenon was observed with other plasmids which confer disadvantage on the carrying strains in gnotobiotic conditions (Duval-Iflah et al. 1981), and in human gut (Duval-Iflah et al. 1982). The presence of complex human digestive flora was shown to have an effect upon the population level of donor and recipient strains and on the kinetics of transconjugant formation, but not on the rate of transfer (Duval-Iflah et al. 1980). *Bacteroides* were shown to inhibit R-factor transfer *in vitro* (Anderson 1975) but not *in vivo* (Duval-Iflah et al. 1980). Antibiotics are known to favour apparition and establishment of transconjugants. The mechanisms involved are probably numerous: antibiotics may promote gut colonization with the donor and /or with transconjugant strains either by disturbing complex flora or by eliminating a specific strain which exerts an intraspecific antagonism upon the transconjugants (Duval-Iflah et al. 1980, 1981). Moreover some data indicate that antibiotics play an additional role in favoring genetic exchange (Brefort et al. 1977; Doucet-Populaire et al. 1991).

Potential mechanisms of gene transfer in the gut.

Most of experimental studies on gene transfer in the gut were mainly concerned with conjugation. However some data indicate that other ways of transfer might occur. Genetic recombinations between *E. coli* K-12 strains have been described (Ducluzeau et al. 1967; Jones and Curtis 1970) and between strains of *C. perfringens* (Brefort et al. 1977). Propagation of staphylococcal phages (Duval-Iflah 1972) as well as ribonucleic phages (Ando et al. 1979) were obtained in gnotobiotic mice associated with appropriate propagating strains. Lysogenic conversion of staphylococcal strains together with recombination between phages harbored by these lysogenic strains were obtained in mice gut (Duval-Iflah 1972). These results suggest that transduction can occur in the gut even though it has not yet been described. The fact that free naked DNA from eucaryotic and prokaryotic cells is rapidly degraded in intestine lumen (Maturin and Curtis 1977; Hoskins 1978) suggests that transformation in the gut is an unlikely event. On the other hand the finding of identical genes in very distinct bacteria as *Campylobacter* spp. and *Enterococcus* spp. which are usual inhabitants of the gut is in favour of horizontal genetic exchange between Gram-positive and Gram-negative bacteria in the gut (Lambert et al. 1985; Trieu-Cuot et al. 1985; Trieu-Cuot and Courvalin 1986). Such exchanges are probably the result of a transformation-like event relayed by transposition (Franke et al. 1981; Trieu-Cuot and Courvalin 1986; Brisson-Noël et al 1988).

Conclusions

All the results obtained in conventional and, more particularly, in gnotobiotic conditons indicate that genetic exchanges between microrganisms in the gut are possible. Recent molecular data on specific genes which have been shown to be shared by phylogenetically remote bacteria are in favour of the assumption that the mechanisms by which these exchanges are performed are probably as numerous and as various as those which are being identified every day in laboratory. These results also confirm the assumption that what happens in laboratory also happens in nature.

References

Anderson ES (1975) Viability of, and transfer of a plasmid from *E. coli* K-12 in human intestine. Nature (London) 255:502-504

Anderson JD (1975) Factors that may prevent transfer of antibiotic resistance between Gram-negative bacteria in the gut. J Med Microbiol 8:83-88

Anderson JD, Gillespie WA, Richmond MH (1973) Chemotherapy and antibiotic-resistance transfer between enterobacteria in the human gastro-intestinal tract. J Med Microbiol 6:461-473

Ando A, Furuse K, Watanabe I (1979) Propagation of ribonucleic acid coliphages in gnotobiotic mice. Appl Environ Microbiol 37:1157-1165

Andremont A, Gerbaud G, Tancrède C, Courvalin P (1985) Plasmid-mediated susceptibility to intestinal antagonisms in *Escherichia coli*. Infect Immun 49:751-755

Brefort G, Magot M, Ionesco H, Sebald M (1977) Characterization and transferability of *Clostridium perfringens* plasmids. Plasmid 1:52-66

Brisson-Noël A, Arthur M, Courvalin P (1988) Evidence for natural gene transfer from Gram-positive cocci to *Escherichia coli*. J Bacteriol 170:1739-1745

Cohen PS, Laux DC (1985) *E.coli* colonization of the mammalian colon : understanding the process. Recomb DNA Tech Bull 8:51-54

Cohen PS, Pilsucki RW, Myhal ML, Rosen CA, Laux DC, Cabelli VJ (1979) Colonization potentials of male and female *E. coli* K-12 strains, *E. coli* B, and human faecal *E. coli* strains in the mouse Gl tract. Recomb DNA Tech Bull 2:106-113

Doucet-Populaire F, Trieu-Cuot P, Dosbaa I, Andremont A, and Courvalin P (1991) Inducible transfer of conjugative transposon Tn*1545* from *Enterococcus faecalis* to *Listeria monocytogenes* in the digestive tracts of gnotobiotic mice. Antimicrob Agents Chemother 35:185-187

Ducluzeau R, Gallinha A (1967) Recombinaison *in vivo* entre une souche Hfr et une souche F⁻ de *Escherichia coli* K-12 encemencées dans le tube digestif de souris axéniques. C R Acad Sci Paris 264:177-179

Duval-Iflah Y (1972) Recombinaison *in vivo* et *in vitro* entre phages de *Staphylococcus pyogenes*. C R Acad Sci Paris 275:3035-3038

Duval-Iflah Y, Gruzza M (1990) Use of germ-free mice in the study of gene transfer from genetically modified lactic bacteria. Commission of the European Communities: Biotechnology Action Programme, Sectorial Meeting on Risk Assessment, Padova, Italy

Duval-Iflah Y, Ouriet MF, Moreau C, Daniel N, Gabillan JC Raibaud P (1982) Implantation précoce d'une souche de *Escherichia coli* dans l'intestin de nouveaux-nés humains: effet de barrière vis-à-vis de souches de *E. coli* antibiorésistants. Ann Microbiol (Institut Pasteur) 133 A:393-408

Duval-Iflah Y, Raibaud P, Rousseau M (1981) Antagonisms among isogenic strains of *Escherichia coli* in the digestive tracts of gnotobiotic mice. Infect Immun 34:957-969

Duval-Iflah Y, Raibaud P, Tancrède C, Rousseau M (1980) R-plasmid transfer from *Serratia liquefaciens* to *Escherichia coli in vitro* and *in vivo* in the digestive tract of gnotobiotic mice associated with human faecal flora. Infect Immun 28:981-990

Farrar WE, Jr, Edison M, Guerry P, Falkow S, Drusin LM, Roberts RB (1972) Interbacterial transfer of R factor in the human intestine : *In vivo* acquisition of R-factor-mediated kanamycin resistance by a multiresistant strain of *Shigella sonnei*. J Infect Dis 126:27-33

Franke AE, Clewell DB (1981) Evidence for conjugal transfer of a *Streptococcus faecalis* transposon (Tn*916*) in the absence of plasmid DNA. Cold Spring Harbor Symp Quant Biol 45:77-80

Freter R, Freter RR, Brickner H (1983) Experimental and mathematical models of *Escherichia coli* plasmid transfer *in vitro* and *in vivo*. Infect Immun 39:60-84

Gruzza M, Duval-Iflah Y, Ducluzeau (1990) *In vivo* establishment of genetically engineered *Lactococci* in gnotobiotic mice; plasmid transfer to *Enterococcs faecalis*. Proceedings of the 10th International Symposium on Gnotobiology, Leiden, The Netherlands. In press

Guinée PAM (1965) Transfer of multiple drug resistance from *Escherichia coli* to *Salmonella typhimurium* in the mouse intestine. Antonie van Leeuwenhoek 31:314-322

Guinée PAM (1968) R transfer to *S. Panama in vitro* and *in vivo*. Antonie van Leeuwenhoek 34:93-98

Guinée PAM (1970) Resistance transfer to the resident intestinal *Escherichia coli* of rats. J Microbiol 102:291-292

Gyles C, Falkow S, Rollins L (1978) *In vivo* transfer of an *Escherichia coli* enterotoxin plasmid possessing genes for drug resistance. Am J Vet Res 39:1438-1441

Hoskins LC (1978) Host and microbial DNA in the gut lumen. J Infect Dis 137:694-703

Jones RT, Curtis R, III (1970) Genetic exchange between *Escherichia coli* strains in the mouse intestine. J Bacteriol 103:71-80

Kasuya M (1964) Transfer of drug resistance between enteric bacteria induced in the mouse intestine. J Bacteriol 88:322-328

Lambert T, Gerbaud G, Trieu-Cuot P, Courvalin P (1985) Structural relationship between the genes encoding 3'-aminoglycoside phosphotransferases in *Campylobacter* and in Gram-positive cocci. Ann Microbiol (Institut Pasteur) 136B:135-150

Laux DC, Cabelli VJ, Cohen PS (1982) The effect of plasmid gene expression on the colonizing ability of *E.coli* HS in mice. Recomb DNA Tech Bull 5:1-5

Levy SB (1984) Survival of plasmids in *Escherichia coli*. In "Genetic Manipulation : Impact on Man and Society" (Arber W, Illmensee K, Peacock J, Starlinger P, eds), pp 19-28 ICSU Press, Paris

Levy SB, Marshall B (1979) Survival of *E. coli* host-vector systems in the human intestinal tract. Recomb DNA Tech Bull 2:77-80

Levy SB, Marshall B, Rowse-Eagle D (1980) Survival of *Escherichia coli* host-vector systems in the mammalian intestine. Science 209:391-394

Marshall B, Schluederberg S, Tachibana C, Levy SB (1981) Survival and transfer in the human gut of poorly mobilizable (pBR322) and of transferable plasmids from the same carrier *E.coli*. Gene 14:145-154

Maturin L, Sr, and Curtiss R, III (1977) Degradation of DNA by nucleases in intestinal tract of rats. Science 196:216-218

Morelli L, Sarra PG, Bottazzi V (1988) *In vivo* transfer of pAMß1 from *Lactococcus reuteri* to *Enterococcus faecalis*. J Appl Bacteriol 65:371-375

Myhal ML, Laux DC, Cohen PS (1982) Relative colonizing abilities of human faecal and K-12 strains of *Escherichia coli* in the large intestines of streptomycin-treated mice. Eur J Clin Microbiol 1:186-192

Petrocheilou V, Grinsted J, Richmond MH (1976) R-plasmid transfer *in vivo* in the absence of antibiotic selection pressure. Antimicrob Agents Chemother 10:753-761

Sansonetti P, Lafont JP, Jaffé-Brachet A, Guillot JF, Chaslus-Dancla E (1980) Parameters controlling interbacterial plasmid spreading in a gnotobiotic chicken gut system: influence of plasmid and bacterial mutations. Antimicrob Agents Chemother 17:327-333

Sears HJ, Brownlee I, Uchiyama JK (1950) Persistence of individual strains of *Escherichia coli* in the intestinal tract of man. J Bacteriol 59:293-301

Smith C, Jr, Milewski E, Martin MA (1985) The effect of colonizing mice with laboratory and wild type strains of *E. coli* containing tumor virus genomes. Recomb DNA Tech Bull 8:47-51

Smith HW (1969) Transfer of antibiotic resistance from animal and human strains of *Escherichia coli* to resident *E. coli* in the alimentary tract of man. Lancet 1:1174-1176.

Smith HW, (1971) Observations on *in vivo* transfer of R-factors. Ann NY Acad Sci 182:80-90

Smith HW (1975) Survival of orally administered *E.coli* K-12 in alimentary tract of man. Nature (London) 255:500-502

Trieu-Cuot P, Courvalin P (1986) Evolution and transfer of aminoglycoside resistance genes under natural conditions. J Antimicrob Chemother 18:93-102

Trieu-Cuot P, Gerbaud G, Lambert T, Courvalin P (1985) *In vivo* transfer of genetic information between Gram-positive and Gram-negative bacteria. The EMBO J 4:3583-3587

Wells CL, Johnson WJ, Kan CM, Balish E (1978) Inhability of debilitated *Escherichia coli* X1776 to colonize germ-free rodents. Nature (London) 274:397

Williams PH (1977) Plasmid tranfer in the human alimentary tract. FEMS Microbiol Lett 2:91-95

Conjugal Transfer of Genetic Information in Gnotobiotic Mice.

F. Doucet-Populaire
Laboratoire de Bactériologie
Faculté de Médecine Pitié-Salpêtrière
91 Bd de l'Hôpital
75634, Paris Cedex 13
France

Dissemination of antibiotic resistance determinants has greatly increased over the past two decades, and its spread is undoubtedly a response to the increasing use of antimicrobial agents. Antibiotic resistance genes therefore constitute a suitable system to study the extent of horizontal gene transfer among prokaryotes. In a attempt to elucidate the mechanism of *in vivo* dissemination of genetic information between phylogenetically remote organisms, we have studied the transfer of two different genetic elements in an animal model. Since the intestinal ecosystem is the most probable meeting point for the bacterial species studied in nature, donors and recipients were associated for 35 or 40 days in the digestive tract of gnotobiotic mice. The transfer was conducted in groups of 6 adult germ-free C3H mice (Centre de selection des animaux de laboratoire, Orléans, France). The mice were maintained in Trexler-type isolators. Animals were fed *at libitum* with a commercial diet sterilized by gamma irradiation. They were supplied with autoclaved drinking water acidified to pH 3 to prevent any bacterial growth in the beverage. After 12 hours without water, germ-free mice were inoculated intragastrically with 1 ml of a broth culture containing 10^8 colony forming units (CFU) of the recipient strain. We did the same inoculation 8 days later with 10^8 CFU of the donor strain harboring the movable genetic element studied (Andremont et al.

1985). Faecal samples were collected directly at the anus of the mice 5 times a week to monitor the level of the bacterial populations (Freter et al. 1983). The freshly passed pellets were weighed, homogenized and serially diluted in saline, and 0.1-ml fractions were plated onto selective media to enumerate donors, recipients and transconjugants. When necessary, the mice were killed, the intestinal tract was removed from pylorus to rectum, weighed, diluted 10-fold in saline, homogenized with an Ultraturax mixer, and the different populations were enumerated.

We have studied in this model, the transfer of the conjugative transposon Tn*1545*. This 25.3 kb streptococcal element confers resistance to kanamycin, erythromycin and tetracycline and is self transferable to a large variety of Gram-positive bacteria (Courvalin et al. 1986). The conjugal transfer of Tn*1545* from *Enterococcus faecalis* to *Listeria monocytogenes* was tested *in vitro* by mating on solid medium (Doucet-Populaire et al. 1991), and *in vivo* in two groups of 6 adults germ-free mice. The transconjugants were selected on their kanamycin resistance. One group of mice received drinking water supplemented with 1 µl/ml of tetracycline.

Strains of *L. monocytogenes* resistant to kanamycin were isolated from the feces of each mouse at concentration ranging from 2 to 3.3 log10 CFU/g of feces the day following inoculation of the donor strain and from all faecal samples collected thereafter. *In vitro*, in 3 independent experiments, transfer of Tn*1545* was obtained at an average frequency of 2.5×10^{-7}. Presence of tetracycline resulted in a 20-fold increase of transfer frequency. Comparison of bacterial counts in the feces at the end of experiment showed that the presence of tetracycline in the drinking water resulted in a 10-fold increase in the *in vivo* transfer frequency of Tn*1545* (Table 1).

These results indicate that streptococcal transposon Tn*1545* can be conjugatively transferred from *E. faecalis* to *L. monocytogenes* in the gastrointestinal tract of gnotobiotic mice although a similar intergeneric transfer occured at low frequency *in vitro* and that low doses of tetracycline increase the transfer by a yet unknown mechanism.

Table 1. Transfer of Tn*1545* from *Enterococcus faecalis* to *Listeria monocytogenes* [a]

	Transfer frequency
***in vitro* transfer**[b]	
Without Tc in the mating medium.	2.5×10^{-7}
With Tc (0.2 mg/l) in the mating medium.	5.6×10^{-6}
***in vivo* transfer**[c]	
Gnotobiotic mice supplied with pure drinking water.	1.1×10^{-8}
Gnotobiotic mice supplied with drinking water supplemented with Tc (1mg/l).	1.3×10^{-7}

[a] The selection medium was brain heart infusion agar containing rifampicin (40 mg/l), fucidic acid (20 mg/l), and kanamycin (20 mg/l). The co-transfer of the two other resistance markers of Tn*1545* was determined by replica plating. Tc, tetracycline.
[b] Transfer frequencies are expressed as the number of transconjugants per donor colony formed after the mating period and values are mean of three independant matings.
[c] Transfer frequencies are expressed as the number of transconjugants per donor cells extracted from the intestine of the gnotobiotic mice sacrificed after 35 days of experiment.

The transfer of the conjugative shuttle plasmid pAT191 that confers resistance to kanamycin was studied from *E. faecalis* to *Escherichia coli*. Plasmid pAT191 was previously described (Trieu-Cuot et al. 1988) and is composed as follows. This 32.5-kb chimeric vector contains the origin of replication and the transfer functions of the broad-host-range enterococcal plasmid pAMß1, the origin of replication of pBR322, and a kanamycin resistance gene (*aph*A-3). This shuttle vector is self-transferable by conjugation on solid medium from *E. faecalis* to *E. coli* with an average frequency of 5×10^{-9} per donor colony formed after the mating period (Trieu-Cuot et al. 1988). Colonies of *E. coli* resistant to kanamycin were isolated from the feces of two mice, respectively, on day 25 and 35 after the beginning of the experiment and never thereafter (Fig. 1).

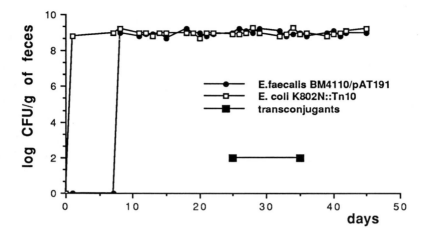

Fig. 1. Bacterial counts in the feces of gnotobiotic mice. Germ-free mice were inoculated on day zero with 10^8 recipient cells (*E. coli* K802N::Tn*10*) and on day 8 with 10^8 donor cells (*E. faecalis* BM4110 harboring pAT191). Data are mean $\log_{10}$ counts per gram of feces.

The transfer frequency of pAT191, expressed as the number of transconjugants per donor cell isolated from intestines of sacrificed mice, was 3×10^{-9}. The two transconjugants were found to contain 6.4-kb plasmids. The presence of *aph*A-3, the kanamycin resistance gene carried by pAT191, in these plasmids was confirmed by DNA-DNA hybridization with a specific probe. The two plasmids were indistinguishable on the basis of their restriction profiles. The genetic transfer in the feces recovery growth medium was found to be $\leq 10^{-10}$ (Doucet-Populaire et al. 1992). These results indicate that DNA transfer takes place between Gram-positive and Gram-negative bacteria in the digestive tract of gnotobiotic mice.

Enterococci and Streptococci harboring resistance plasmids or transposons are common in the digestive tract of humans and animals. It has been proposed that these bacteria might serve as a reservoir of resistance genes, not only for other Gram-positive bacteria as *Listeria* but also for Gram-negative organisms (Trieu-Cuot et al. 1987). We have shown in this study that genetic information can be transferred from *E. faecalis* to *L. monocytogenes* and to *E. coli* in the gastrointestinal tract of gnotobiotic mice. However, this animal model is probably more favorable for genetic exchange by conjugation since it allows intestinal colonization by donors and recipients in counts greater than those observed in natural conditions. It suggests that conjugation is therefore a mechanism that could account for resistance gene flux.

References

Andremont A, Gerbaud G, Tancrede C, Courvalin P (1985) Plasmid-mediated susceptibility to intestinal microbial antagonisms in *Escherichia coli*. Infect Immun 49:751-755

Courvalin P, Carlier C (1986) Transposable multiple antibiotic resistance in *Streptococcus pneumoniae*. Mol Gen Genet 205:291-297

Doucet-Populaire F, Trieu-Cuot P, Dosbaa I, Andremont A, Courvalin P (1991) Inducible transfer of conjugative transposon Tn*1545* from *Enterococcus faecalis* to *Listeria monocytogenes* in the digestive tracts of gnotobiotic mice. Antimicrob Agents Chemother 35:185-187

Doucet-Populaire F, Trieu-Cuot P, Andremont A, Courvalin P (1992) Conjugal transfer of plasmid DNA from *Enterococcus faecalis* to *Escherichia coli* in the digestive tracts of gnotobiotic mice. Antimicrob Agents Chemother 36:502-504.

Freter R, Brickner H, Fekete J, Vickerman MM, Carcey KE (1983) Survival and implantation of *Escherichia coli* in the intestinal tract. Infect Immun 39:686-703.

Trieu-Cuot P, Carlier C, Courvalin P (1988) Conjugative plasmid transfer from *Enterococcus faecalis* to *Escherichia coli*. J Bacteriol 170:4388-4391.

Trieu-Cuot P, Arthur M, Courvalin P (1987) Transfer of genetic information between Gram-positive and Gram-negative under natural conditions, p.65-68. In R Curtiss III and JJ Feretti (ed.), Streptococcal genetics. American Society for Microbiology, Washington D C

In vivo Transfer of a Conjugative Plasmid between Isogenic *Escherichia coli* Strains in the Gut of Chickens, in the Presence and Absence of Selective Pressure

J.F. Guillot[1,2], J.L. Boucaud[2]
Université[1] et INRA[2]
Laboratoire de Microbiologie
29, rue du Pont Volant
37023, Tours Cedex
France

Introduction

The faecal flora of poultry and other farm animals includes a high proportion of antibiotic-resistant bacteria. Among these intestinal bacteria, the enterobacteria harbor frequently resistance plasmids and they are able to colonize the gut of chickens without any selective pressure (Guillot et al. 1977).

Conjugative transfer is a frequent property of enterobacterial R plasmids. This transfer has been observed *in vitro* and also in natural conditions, particularly in the gut of animals and humans (Arthur et al. 1987; Duval-Iflah et al. 1980; Freter et al. 1983; Guillot and Boucaud 1988; Mc Connell et al. 1991; Sansonetti et al. 1980; Shimoda et al. 1985). The variety of bacterial species present in gut, mixing, and high bacterial density may be favorable conditions for conjugative transfer.

In this work, we have studied the transfer of a natural plasmid between two mutants of the same *E. coli* strain of avian origin, in the gut of dixenic and gnotobiotic chickens, treated or not with tetracycline.

Materials and Methods

"In vitro" studies

- Bacterial strains : The strain BN118 used is an antibiotic sensitive *E. coli*, without extrachromosomal DNA, isolated from the faecal flora of poultry. Two spontaneous chromosomal mutants resistant to nalidixic acid or streptomycin were selected on Szybalski's gradients (Table 1).

- Plasmid : pGB99 is a transferable plasmid coding for chloramphenicol, tetracycline, sulphonamide and trimethoprim resistance, originating from an avian *E.coli* strain. This plasmid has been transferred by conjugation in strain BN118 nal to obtain the original donor strain.

Table 1 : *E. coli* strains

	Strain	Origin	Plasmid	Antibiotic resistance[a]	
				chromosomic	plasmidic
Donor	BN 118	Poultry	pGB99	Nal	CmTcSuTp
Recipient	BN 118	"	/	Str	/

[a] Nal, nalidixic acid; Str, streptomycin; Cm, chloramphenicol; Tc, tetracycline; Su, sulphonamide; Tp, trimethoprim

"In vivo" studies

- Animals : White Leghorn chickens, strain PA12, from the poultry farm of the Station were used (Guillot and Boucaud 1988). The experimental birds were hatched under sterile conditions, and the axenic chicks were reared in sterile, germ-free isolators, in group of ten animals. Sterile water and sterile food deprived of antibiotic supplementation were given. The germ-free status of experimental animals was checked by incubating their droppings aerobically and anaerobically before experimental contamination.

- Inoculation of animals : A 0.5 ml amount of an overnight culture of recipient and donor bacteria (10 /ml) was successively force fed to 1 week-old axenic chickens, using an oesophageal catheter mounted on a syringe. In the two groups of gnotoxenic chickens the animals were inoculated just after being administered the recipient strain with an avian microflora harboring only antibiotic sensitive *E. coli*.

- Antibiotic treatment : Two groups of birds were dosed with antibiotics. Tetracycline was added to the drinking water (0.5 g/l) for 6 days beginning one day after the donor strain was inoculated.

- Faeces sampling : Three animals out of ten were regularly sampled in each isolator. Their faeces were collected from the cloaca into separate sterile tubes and maintained at 4°C to prevent plasmid transfer after sampling. The tubes were weighed and appropriate dilutions were plated on selective media for donor, recipient and transconjugant numeration.

Results

"In vivo" transfer of pGB99 CmTcSuTp in dixenic chickens
In the absence of selective pressure, plasmid transfer occurred rapidly after the colonization of the gut and the transconjugants persisted stably (Fig. 1).

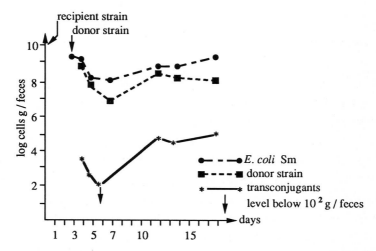

Fig. 1 *In vivo* transfer of pGB99 CmTcSuTp in dixenic chickens in the absence of selective prressure

Tetracycline selected the donor strain and the transconjugants, and the recipient strain could not be recovered from the faeces after the first week of antibiotic use (Fig. 2).

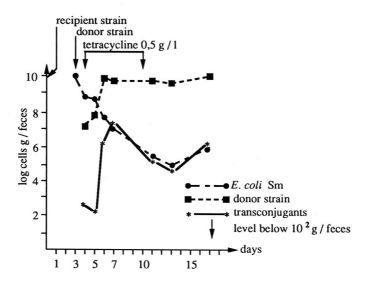

Fig. 2. *In vivo* transfer of pGB99 CmTcSuTp in dixenic chickens in the presence of selective pressure

"In vivo" transfer of pGB99 CmTcSuTp in gnotoxenic chickens
The plasmid transfer was not observed between the two mutants of *E. coli* in the gut of birds in the absence of tetracycline (Fig. 3). After 14 days, the population of Cm resistant *E. coli* was inferior to the population of Sm resistant *E. coli* and superior to the population of donor strain and presumably corresponded to transconjugants between the donor strain and resident *E. coli* of the flora.

In the gut of birds receiving tetracycline, transconjugants of BN 118 were obtained just at the end of the period of treatment. Transconjugants between the donor strain and the resident *E. coli* seems to have also occurred. These transconjugants increased in the microflora during therapy and persisted at a dominant level after the end of the treatment (Fig. 4).

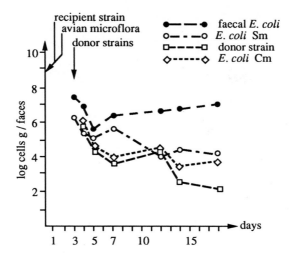

Fig. 3. *In vivo* transfer of pGB99 CmTcSuTp in gnotoxenic chickens in the absence of selective pressure

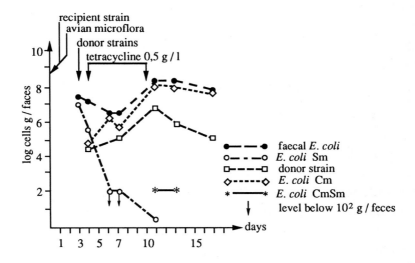

Fig. 4. *In vivo* transfer of pGB99 CmTcSuTp in gnotoxenic chickens in the presence of selective pressure

Discussion

Our results confirm that an *in vivo* transfer of plasmids occurs in the gut of chickens. This transfer occurs rapidly in dixenic animals, irrespective of selective pressure, and the transfer rate appeared higher in dixenic than in gnotobiotic animals. The proportion of transconjugants obtained was smaller than that reported by other authors (Duval-Iflah et al. 1980; Sansonetti et al. 1980; Shimoda et al. 1985). Several factors : bacterial strains, repression of transfer in wild-type plasmids...could account for these results (Sansonetti et al. 1980).

The other components of the avian microflora repressed the transfer both directly through the presence of a dense and complex bacterial population and indirectly through the physico-chemical conditions created in the intestines.

The use of antibiotics selected the resistant strains, donor strains and transconjugants, including those resulting from transfer to the resident *E. coli,* but it is difficult to assume that those substances act on the transfer rate itself.

In field conditions the results obtained in dixenic animals could correspond to the situation in newborn chickens and those from gnotobiotic animals to the intestinal conditions prevailing in older animals.

References

Arthur M, Brisson-Noel A, Courvalin P (1987) Origin and evolution of genes specifying resistance to macrolide, lincosamide and streptogramin antibiotics, data and hypotheses. J Antimicrob Chemother 20:783-802.

Duval-Iflah Y, Raibaud P, Tancrede C, Rousseau M (1980) R-plasmid transfer from *Serratia liquefaciens* to *Escherichia coli in vitro* and *in vivo* in the digestive tract of gnotobiotic mice associated with human fecal flora. Infect Immun 28:981-990.

Freter R, Freter R, Brickner H (1983) Experimental and mathematical models of *E. coli* plasmid transfer *in vitro* and *in vivo*. Infect Immun 39:60-84.

Guillot JF, Boucaud JL (1988). Transfert plasmidique *in vivo* entre souches d'*Escherichia coli* dans le tube digestif du poulet. Pathol Biol 36:655-659.

Guillot JF, Chaslus-Dangla E, Lafont JP (1977) Spontaneous implantation of antibiotic-resistant enterobacteriaceae in the digestive tract of chickens in the absence of selective pressure. Antimicrob Agents Chemother 12:697-702.

McConnell MA, Mercer AA, Tannock GW (1991) Transfer of plasmid pAMß1 between members of the normal microflora inhabiting the murine digestive tract and modification of the plasmid in a *Lactobacillus reuteri* host. Microb Ecol Health Disease 4:343-355.

Sansonetti P, Lafont JP, Jaffe-Brachet A, Guillot JF, Chaslus-Dancla E (1980) Parameters controlling interbacterial spreading in a gnotoxenic chicken gut system: influence of plasmid and bacterial mutations. Antimicrob Agents Chemother 17:327-333.

Shimoda K, Maejima K, Terakado N (1985) *In vivo* conversion of transferable plasmid into R-plasmid in the intestine of gnotobiotic mice. Jpn J Vet Sci 47:669-672.

Origin and Spread of Penicillin-Resistance in *Neisseria meningitidis*

J. Campos, M.C. Fusté, J. Vazquez[1], J.A. Saez-Nieto[1] and M. Viñas.
Department of Microbiology and Parasitology
University of Barcelona
08028 Barcelona
[1] National Center of Microbiology.
Madrid.
Spain.

Introduction

Moderate penicillin resistance in the human pathogen *Neisseria meningitidis* emerged seven years ago in Spain (Saez-Nieto and Campos 1988). Since the first isolate in 1985, the frequency of Pen[R] strain isolation is increasing especially in the area surrounding Barcelona (reaching 40% in 1991).

Mechanism of penicillin resistance.

The molecular mechanism involved in penicillin resistance in *N. meningitidis* has been studied by different groups using several methodologies. It has been pointed out that no enzymatic degradation of the antibiotics could be detected in any of the strains studied (Mendelman et al. 1988). On the other hand, when whole-cell were labeled with [3H]-penicillin G (0.006 µg/100 µl) and the radioactive label was compared with those obtained in susceptible strains, three PBPs (penicillin binding proteins)with similar mobilities were detected in all strains studied. The PBP 2 of the pen[R] strains bound less [^{3}H]-penicillin than those corresponding to sensitive isolates (Mendelman et al. 1989). The loss of affinity for penicillin of PBP2 was thus considered the primary cause of relative penicillin resistance in *N. meningitidis* (Mendelman et al.1988).

Origin and interest of penicillin resistance in meningococci

Because penicillin G is the drug of choice to treat meningococcal meningitis as well as meningococcal septicaemia, the emergence of a loss of susceptibility to penicillin may eventually be a handicap in therapeutics. Initially, the resistant strains exhibited increased MIC's (minimal inhibitory concentrations) against penicillin, but they were still susceptible from a clinical point of view, although the levels of resistance recently detected in the new isolates reach the penicillin concentrations expected in the infected tissues (Saez-Nieto et al. 1992). The origin of the first strains exhibiting decreased susceptibility to penicillin is unknown. Saez-Nieto et al.(1990) studied the ability of meningococcus to acquire genetic information from different donors, leading to resistant phenotypes. They used *N. polysaccharea, N. gonorrhoeae* and *N. lactamica* as donor strains; these species presented a moderate resistance to penicillin. Since the meningococci are naturally transformable, chromosomal DNA was obtained and used in transformation experiments. Authors were able to demonstrate the transformation of a type strain of *N. meningitidis* to low level penicillin resistance with DNA from both *N. polysaccharea* and *N. lactamica*. For these reasons, the authors mentioned suggested that the commensal *Neisseriaceae* could be a reservoir of ß-lactam antibiotic resistance in both commensal and pathogenic *Neisseria*.

Population genetics of Pen^R meningococci

The emergence of a chromosomally encoded character and its rapid spread suggested a mutation, followed by a dispersion of the clone presenting the selective advantage. However, data concerning the ability of *Neisseria* sp. to acquire penicillin resistance through transformation suggested the possible role of other mechanisms of spread. Recently, multilocus isoenzyme electrophoresis has been established as a powerful technique for genetic population studies. The simultaneous treatment of data from enzymatic polymorphism, serogrouping-serotyping, hybridization with specific probes for the *penA* gene, and restriction fragment length polymorphism (RFLP) of the gene have been used in order to establish the genetic origin and spread of the penicillin-resistant clones (Campos et al. 1992). Data are summarized in Table 1 :

Table 1. Characteristics of *Neisseria* sp. strains

Strain	Serogroup-serotype	MIC penicillin	ampicillin	RFLP class	N.fla. probe
1116	B 15P1.1.16	0.025	0.025	S	N
1092	B 15P1.1.16	0.025	0.025	S	N
1154	B 15P1.15	0.025	0.025	S	N
1059	B 15P1.16	0.025	0.05	S	N
1094	B 2b	0.025	0.05	S	N
1124	B 2b	0.025	0.025	S	N
1151	B 5P1.16	0.025	0.025	S	N
1107	C 2b	0.2	0.3	C^2	P
1148	C 2b	0.2	0.2	C^2	P
1103	C 2b	0.2	0.3	C^2	P
1118	C 2b	0.2	0.3	C^2	P
1106	C 2b	0.2	0.3	C^2	P
1091	C 2b	0.2	0.2	C^2	P
1046	B 2b	0.2	0.2	C^2	P
1155	B 15P1.1,15	0.1	0.1	C^2	P
1085	C 2b	0.2	0.2	C^2	P
1058	B 2P1.15	0.1	0.2	B^I	N
1083	B15P1.14,15	0.1	0.3	B^I	N
1143	B 4P1.15	0.1	0.2	B^I	N
1061	B 4P1.15	0.1	0.2	B^I	N
1042	B 15P1.16	0.1	0.2	B^I	N
1123	B 4P1.15	0.2	0.3	B^I	N

Table 1. Continued

1157	B 4P1.1,7	0.1	0.2	B^I	N
1033	B NT	0.1	0.2	B^I	N
1045	B15P1.17,16	0.05	0.2	B^I	N
1146	B5P1.16	0.1	0.2	B^I	N
1066	B 4P1.15	0.05	0.1	B^{II}	N
1129	B 4P1.15	0.5	0.8	B^{II}	N
1079	B 4P1.15	0.1	0.2	B^{II}	N
1159	B 15P1.16	0.1	0.2	B^{II}	N
1068	B 4P1.15	0.1	0.1	B^{II}	N
1099	B 4P1.15	0.05	0.1	B^{II}	N
1072	B 15P1.16	0.05	0.2	B^{II}	N
1019	B 4P1.15	0.1	0.2	B^{III}	P
1039	B 14P1.1,7	0.1	0.4	B^{III}	P
1037	B 14P1.1,7	0.2	0.3	B^{III}	P
1060	B 14P1.1,7	0.2	0.3	B^{III}	P
1080	C 2b	0.2	0.3	V	P
1077	B NT	0.1	0.2	V	P
1047	C 2b	0.2	0.4	V	?
1053	B 15P1.7,16	0.2	0.2	V	P
1056	B 15P1.7,16	0.1	0.2	V	P
1082	B 1:-	0.1	0.2	V	P
1113	B NT	0.05	0.2	V	P
1031	B 15P1.7,16	0.1	0.3	V	P
1095	B 15P1.16	0.05	0.2	V	N
1144	B 2b	0.1	0.2	V	P
1149	B 15P1.1,16	0.2	0.2	V	P
1158	B 15P1.2	0.2	0.2	V	P

(P) Positive; (N) Negative; (?) signal too weak; (Str.) strain. S, C^2, B^I, B^{II}, B^{III} and V are different RFLP patterns.

High genetic diversity (H=0,6) was observed when the electrophoretic mobilities of various chromosomally encoded enzymes were studied. These results, together with those corresponding to serogrouping/serotyping and RFLP patterns, demonstrate that penicillin resistant populations of meningococci have not arisen from the spread of one resistant clone, but from several clones at least. However, taking into account the special characteristics of the species, the role of genetic exchange cannot be ruled out. It has been pointed out that *Neisseria* could acquire sequences by interstrain transformation following uptake of chromosomal DNA released by other neisserial bacteria. This mechanism has proved to be crucial in the modification of several surface virulence factors. As pointed out by Saez-Nieto et al. (1990), the commensal *Neisseriaceae* (naturally penicillin resistant) could be a reservoir of ß-lactam antibiotic resistance. Meningococcus could acquire the sequences encoding resistance through transformation. Although is easy to transform *Neisseria* in laboratory conditions, this remains to be demonstrated in other conditions. Further work is needed in order to evaluate the hypothesis. Perhaps the use of mesocosms, in which the conditions of oral cavity and human nasopharynx could be mimicked, could be of interest to evaluate the efficiency of genetic exchange between commensal species and meningococci in their natural habitat.

Acknowledgements

This research was supported by grants 88/1483 (FIS) and PM 1097-c03-02 (DGICYT).

References

Campos J, Fusté MC, Trujillo C, Saez-Nieto JA, Vazquez J, Lorén JG, Viñas M, Spratt BG (1992) Genetic Diversity of penicillin-resistant *Neisseria meningitidis* J Infect Dis (in press)

Mendelman PM, Campos J, Chaffin DO, Serfass DA, Smith AL, Saez-Nieto JA (1988) Relative penicillin G resistance in *Neisseria meningitidis* and reduced affinity of penicillin-binding protein 3. Antimicrob Agents Chemother 32:706-709

Mendelman PM, Caugant DA, Kalaitzoglou G, Wedege E, Chaffin DO, Campos J, Saez-Nieto JA, Viñas M, Selander RK (1989) Genetic diversity of penicillin G-resistant *Neisseria meningitidis* in Spain. Infect Immun 57:1025-1029.

Saez-Nieto JA, Campos J (1988) Penicillin-resistant strains of *Neisseria meningitidis* in Spain. Lancet 1:1452-1453

Saez-Nieto JA, Lujan R, Martinez-Suarez JV, Berron S, Vazquez JA, Viñas M, Campos J (1990) *Neisseria lactamica* and *Neisseria polysaccharea* as possible source of meningococcal ß-lactam resistance by genetic transformation. Antimicrob Agents Chemother 34:2269-2272

Saez-Nieto JA, Lujan R, Berron S, Campos J, Viñas M, Fusté C, Vazquez JA, Zhang QY, Bowler LD, Martin-Suarez JV, Spratt BG (1992) Epidemiology and molecular basis of penicillin-resistant *Neisseria meningitidis* in Spain. Clin Infect Dis 14:394-402

Section 5

GENETICALLY ENGINEERED MICROORGANISMS

Release of Genetically Modified Microorganisms in Natural Environments : Scientific and Ethical Problems

A. Klier
Institut Pasteur
25 rue du Dr Roux
75724 Paris Cedex 15
France

Genetically engineered organisms are likely to be the cornerstone of the commercial application of biotechnology in the coming decades. The exponential increase of fundamental understanding in molecular biology over the last 20 years, is such that foreign DNA can be inserted into the genome of most organisms : it is now possible to introduce new and useful genes into organisms or to inactivate or modify genes within organisms, thereby removing certain traits or disarming pathogens. A wealth of potential applications in all fields including agriculture, medicine, chemistry has been revealed and several new pharmaceutical products are now on the market, allowing revolutionary new approaches to therapy and prophylaxis. In addition, herbicide-resistant and insecticide-producing plants have been prepared and have been shown to be effective in controlled trials. However, most of these applications are not yet widely used, since there is an increased awareness of the need to assess the possible consequences of the controlled release of genetically modified organisms into the environment. Moreover, the new possibilities of increasing the insect resistance of plants or improving their growth rate by the deliberate release of genetically modified microorganisms has caused concern among some groups and individuals.

The predetermination of the consequences of such a release is often termed "risk assessment". However, since no one knows what the risk might be, a more appropriate term would be "consequence assessment". This can be positive, neutral or negative with respect to the effects on man and his environment. Moreover, the balance between the positive and the negative consequences should be carefully evaluated. When a modified microorganism is released for a particular purpose, several outcomes may be envisaged :

1) the microorganism cannot survive and perishes in the environment, due to some disadvantages (such as the presence of plasmid or the requirement for specific nutrients),
2) the microorganism establishes itself in the manner anticipated by its use and performs its desired function,
3) the microorganism becomes a "rascal" and when released, affects the environment in an unforeseen and deleterious manner. It could do this in several different ways :
   - phenotypically (expression of unwanted growth or competition characteristics),
   - genetically (transfer of some or all of the genetic material of the microorganism to indigenous organisms, thereby changing the properties of one or more of the species in the environment in a unwanted way),
   - "others", these have been a number of other suggestions, many of them far fetched.

It is obviously too easy to build frightening scenarios resulting from the presence of unseen recombinant microorganisms in the environment and much of the opposition to the release of these organisms has been based on shrewd misinformation supplied to the public at large, by describing the worst possible eventualities. Many (and probably most) of them have little relationship to good scientific principles, but make excellent subjects for horror movies. Moreover, most of the misinformation comes from people with opportunistic political goals.

It is therefore important that scientists try to establish rational situations that involve the use of genetically engineered organisms. Although some questions can easily be solved by a good basic knowledge of the microorganism, some cannot be resolved without actual field testing, especially those concerning the potential for natural gene transfer and its consequences. Is there measurable gene flux in nature and to what extent may a genetically modified microorganism modify it ? Are some genes, or traits likely to be more troublesome than others, if transfer into indigenous organisms takes place ? It is easy to envisage that the probability of gene transfer depends on a number of unrelated factors, including the nature of the recombinant DNA itself, the host, the selective advantage of the recipient and a multitude of genetic, physiochemical and environmental factors that all need to be tested by experimentation and not by projection from belief.

There is evidence that genes can move between and function in different microbial species. One of the most often cited examples for gene exchange accross wide phylogenic boundaries in nature is the presence of almost identical antibiotic resistance genes in a wide variety of microbes. There is even circumstantial evidence that trans-kingdom transfer has occurred, as judged by DNA sequence similarity between fungal and bacterial β-lactam antibiotic genes (Penalva et al. 1990). If one presumes therefore that gene transfer does occur in the environment, then the important questions concern the mechanisms that are generally used and the barriers which limit the efficacy of transfer and gene expression. Several defined mechanisms for DNA exchange have already been identified and studied in laboratory conditions. There is some evidence that they may operate under natural environmental conditions, but with a low efficiency. Whatever the mechanism involved in transferring exogenous DNA, success will largely but not only, depend on the transferred DNA escaping the recipient host's restriction systems, and being passed on to following generations either as part of, or along with the recipient chromosome. Three types of DNA transfer have been described and investigated in depth.

Transformation of bacteria by exogenous DNA has been the favorite model for the introduction of DNA into host strains. Natural transformation, whereby the recipient spontaneously takes up DNA, has been identified in several genera, including *Acitenobacter, Haemophilus, Pneumococcus, Streptococcus, Bacillus, Pseudomonas* and several others (Stewart and Carlson 1986). In *Bacillus* sp., which is the best studied, the competent state develops naturally at a particular stage in the growth cycle, when the levels of inducible competence factors reach a critical level (Dubnau 1991). For the other species, the mechanisms are not as clearly understood. Presumably in the environment during period of bacterial growth, and whatever the molecular mechanism is involved, a significant proportion of cells would be in the competent state and a supply of exogenous DNA would be available from natural death and lysis of cells, both microbial and otherwise. Experiments which have been performed in sterile soils inoculated with two types of labelled *B. subtilis* strains resulted in the rescue of strains carrying the genetic markers from both parental strains (Graham and Istock 1978). The most likely explanation for the genetic exchange was that it resulted from transformation. However, it is unclear how such results with inoculated laboratory strains in sterile conditions relate to the *in situ* situation. Moreover, the efficacy of transformation is often very low and a high density of recipient cells and contacts between DNA and the cell surface are required. This requirement for high density could be a limiting factor in the environment. In addition, transformant DNA needs to be replicated if it is to survive in the host. It must therefore either be able to recombine into the host genome or to carry a compatible replication origin. Until the processes of competence development, exogenous DNA protection from degradation, and replication are understood, it is assumed that natural transformation can occur in the environment. Whether this mechanism has a major role in gene transfer remains an open question.

In bacterial transduction, genetic information is transferred between host and recipient cells mediated by bacteriophages. Bacteriophages are abundant in the environment, where their survival is strictly dependent on suitable hosts. The main role played by bacteriophages is the control of bacterial populations. However, the ability of some of them to transduce DNA suggests that they may contribute to gene flow. Infection with transducing bacteriophages could rapidly lead to dissemination of recombinant DNA. Environmental studies performed with *Pseudomonas* strains and their bacteriophages demonstrated transduction of plasmids in sterile and nonsterile lake water (Morison et al. 1978). However, the transduction frequency was dramatically lower in the nonsterile environment indicating the effect of the natural microbial community. It was also demonstrated that some *Pseudomonas* bacteriophages are able to use several *Pseudomonas* species as hosts (Kell and Warren 1971). Similarly, some *Bacillus thuringiensis* phages are able to infect and to transduce DNA to *Bacillus cereus* and *Bacillus anthracis* (Thorne 1978). In the laboratory, several factors can be manipulated to maximize transfer or to show low frequency transfer. However, data on biological and physicochemical factors in the environment which may affect gene exchange, is virtually nonexistent. As in the case of transformation, the successful gene transfer requires the DNA to be incorporated into the new host, maintained and passed to the following generations. In order to be inherited by new generations, the foreign DNA would have to provide a selective growth advantage to the host. Alternatively, further cycles of transformation and transduction could help spread the DNA. However, at each additional transfer, the success probability is reduced and thus a barrier to widespread gene exchange by these mechanisms is quickly built.

In conjugation, chromosomal or plasmid DNA is transferred directly between a host strain and the recipient. Most of the work undertaken has been carried out in *E. coli* with narrow host range plasmids. However, with broad host range plasmids, which are able to replicate in a range of different bacteria, the potential of DNA transfer is much greater than would be expected by transformation or transduction. Moreover, many plasmids can mobilize other non transferrable plasmids, and move chromosomal DNA sequences between host and recipient cells *via* transposons. Thus, with conjugative transfer, these are many possibilities for dissemination of DNA to several different hosts. However, there is a gap in our knowledge between what it is possible to show in the lab and what happens in the environment. The evidence of conjugative transfer amoung indigenous organisms in nature is circumstantial and is primarily based on the appearance of genetically almost identical antibiotic resistance genes in many different species and genera. In experimental soil microcosms, conjugative transfer of plasmids has been observed between strains of *E. coli*, *Pseudomonas* sp., *Bacillus* sp. and *Streptomyces* sp. (Stotzky 1989). The physicochemical parameters of the soil and the nature of the plasmids exert a strong influence on the transfer efficacy, as does the density of the microbial community. Until recently, it was generally accepted that conjugation was restricted to closely related species or genera. However, work with recombinant shuttle plasmids has proven that genes can be transferred by conjugation between phylogenetically distant organisms (Trieu-Cuot et al. 1987). Examples of inter kingdom transfer of DNA include the system in which the conjugation functions of the broad host range plasmid RSF1010 can mobilize Ti DNA from *Agrobacterium tumefaciens* to plants (Buchanan-Wollaston 1987) and the conjugation of bifunctional plasmids from *E. coli* to the yeast *Saccharomyces cerevisiae* (Heineman and Sprague 1989). Such exotic transfers suggest that the limitation on successful exchange is not determined by incompatibility between cell to cell contact which prevents transfer, but by the ability of the transferred DNA to become incorporated into the new host. A small number of experiments on intergeneric plasmid transfer have been carried out in soil microcosms between *E. coli* and hosts such as *Pseudomonas* sp. (Henshke and Schmidt 1990), *Rhizobium* sp. (Richaume et al. 1989) and *Alcaligenes* sp. (Torsvik et al. 1990). As the survival potential of host and recipient can be vastly different in soil, the frequency of transfer is difficult to assess. Interestingly, the transfer of a plasmid carrying heavy metal resistance determinants between *E. coli* and *Alcaligenes eutrophus* is significantly enhanced in polluted soils compared to non polluted soils (Torsvik et al. 1990).

All the studies mentioned above have been imaginative and have provided important information about the potential of gene transfer. Several barriers control the efficacy of DNA transfer and its maintenance in a new host. One of the first barriers is the difference in the ecological niches of the bacterial partners and the very low probability that the bacteria would both be present in the same place at a high density and in the competent physiological state. The nature of the various mechanisms of gene exchange also imposes barriers to the entry of exogenous genetic material into the cell. A few molecules escape the restriction system either by autonomously replicating too fast to be cleaved or by integration into the host genome. The efficacy of the integration event is dramatically reduced when functional *mut* genes or their equivalents are present and where there is mismatching. Once established in a recipient, the genes to be expressed require that transcriptional and translational signals be recognized by the host machinery. Finally, the accumulation of the foreign gene product must not have a deleterious effect on the survival of the recipient if the gene is to be stably maintained. Considering all the barriers, the probability of expressing a new phenotype after a successful DNA transfer is very low. However, heterogramic transfers have been demonstrated in lab conditions. The question thus arises : does such successful transfer occur in the field ?

Studies with soils and in aquatic habitats are particularly difficult. It is estimated that one gram of soil may contain about 10^8-10^9 microbes and only about 1 % of these can currently be cultured. Much research is needed on the nature of the organisms in soil and on methods for their isolation and characterization (Top et al. 1990). It is now possible, with PCR, to detect the transfer of a DNA sequence from the donor to an indigenous inhabitant at the frequency of 10^{-6}.

Ultimately, the public wants to know if it is safe to release recombinant bacteria into the environment. The answer to this difficult question is rather subjective since no or little information is available on the impact or detectable changes an introduced microorganism will cause to the environment if it persists longer than expected or if its DNA is transferred to indigenous organisms. If a detectable change does occur, is this necessarily dangerous ? Are there any differences between *E. coli* and a Gram-positive bacterium synthesizing the human growth hormone in this respect ? Until more is known about how microbes interact in the environment, there is no need to be overcautious when dealing with unexpected events involving recombinant bacteria. Nevertheless, deliberate release experiments should only be done after rational consideration of the scientific basis of the study with respect to the nature of the microorganism involved, the site of release and the potential beneficial value of the release.

References

Buchanan-Wollaston V, Passiatore JE, Cannon F (1987) The *mob* and *oriT* mobilization functions of a bacterial plasmid promote its transfer to plants. Nature 328:172-175

Dubnau D (1991) Genetic competence in *Bacillus subtilis*. Microbiol Rev 55:395-424

Graham JB, Istock CA (1978) Genetic exchange in *Bacillus subtilis* in soil. Mol Gen Genet 166:287-290

Heineman JA, Sprague GF Jr (1989) Bacterial conjugative plasmids mobilize DNA transfer between bacteria and yeast. Nature 340:205-208

Henshke RB, Schmidt FRJ (1990) Plasmid mobilization from genetically engineered bacteria to members of the indigenous soil microflora *in situ*. Curr Microbiol 20:105-109

Kell RA, Warren RA (1971) Isolation and properties of a bacteriophage lytic to a wide range of *Pseudomonas*. Can J Microbiol 17:677-682

Morison WD, Miller RV, Sayler GS (1978) Frequency of F116 mediated transduction in *Pseudomonas aeroginosa* in fresh water environments. Appl Environ Microbiol 36:724-727

Penalva MA, Moya A, Dopazo J, Ramon D (1990) Sequence of isopenicillin N synthetase genes suggest horizontal transfer genes from prokaryotes to eukaryotes. Proc R Soc London, 241:164-168

Richaume A, Angle JS, Sadowsky MJ (1989) Influence of soil variables on *in situ* plasmid transfer from *E. coli* to *Rhizobium fredii*. Appl Environ Microbiol 55:1730-1735

Stewart GJ, Carlson CA (1986) The biology of natural transformation. Ann Rev Microbiol 40:211-235

Stotzky G (1989) In gene transfer in the environment. Levy SB and Miller RB (Eds), Mac Graw Hill (Environmental Biotechnology), Vol. 6

Thorne CB (1978) Transduction in *Bacillus thuringiensis*. Appl Environ Microbiol 35:1109-1115

Top E, Mergeay M, Springael D, Verstraete W (1990) Gene escape model : transfer of heavy metal resistance genes from *E. coli* to *Alcaligenes eutrophus* on agar plates and soil samples. Appl Environ Microbiol 56:2471-2479

Torsvik V, Goksoyr J, Daae FL (1990) High diversity in DNA of soil bacteria. Appl Environ Microbiol 56:782-786

Trieu-Cuot P, Carlier C, Martin P, Courvalin P (1987). Plasmid transfer by conjugation from *E. coli* to Gram-positive bacteria. Microbiol Lett 48:289-294

Development of Sensitive Techniques Based on DNA Probes and PCR-Amplification, for the Study of the Survival and Genetic Stability of *Lactococcus* spp. in the Environment and the Gastrointestinal Tract

N. Klijn, A.H. Weerkamp and W.M. de Vos
Netherlands Institute for Dairy Research (NIZO)
P.O. Box 20
6710 BA, Ede
The Netherlands

Introduction

Major research efforts are focused on the genetic improvement of micro-organisms used in industrial food fermentations. This includes strains of *Lactococcus lactis*, which are used in the production of fermented milk products such as cheese. To allow the use of genetically modified organisms (GMOs) in the production of fermented food products, the fate of these organisms and their genetic information must be known.

In order to get a good estimation of the possible effects of the release of genetically modified *L. lactis*, we designed a decision-scheme in which the fate of the GMO and its genetic information of the production process leading to release in the environment is evaluated. In particular, the survival and stability of *Lactococcus* spp. in each step is important for the actual effect on the environment.

Every year more than 10^{20} lactococci are produced in the Netherlands only during the production of fermented milk products and eventually released into the environment but nothing is know about their abilities to survive and grow in nature. Although the isolation of *Lactococcus lactis* species outside dairy environment has been reported (see review in: Sandine et al. 1972) nothing is actually known about the wild population of *Lactococcus* species except for some incidental isolations from frozen peas, fish, raw milk, fermented vegetables and the intestinal tract of termites (Schleifer et al. 1985; Schultz and Breznak 1978; Williams et al. 1990).

To allow the detection and the identification of *Lactococcus* species in various environments, we developed sensitive techniques based on DNA probes (Matthews and Kricka 1988) and polymerase chain reaction (PCR) amplification (Saiki et al. 1988). Species-specific DNA probes were designed, based on hypervariable regions of the 16S rRNA and the sensitivity was improved by using PCR-amplification of the variable regions (Klijn et al. 1991). With these techniques we studied the presence of *Lactococcus* species in the environmental samples associated with cattle, and their survival in the waste flow of a cheese factory and in the human intestinal tract.

Materials and Methods

The development and validation of the species specific DNA probes for *Lactococcus lactis* (subsp. *cremoris* and *lactis*), *L. garviae, L. raffinolactis* and *L. plantarum* based on the variable sequence in the V1 region of the 16S rRNA in combination with PCR amplification of the V1 region have been described by Klijn et al. (1991)

Samples taken from a dairy farm, a bull raising farm and a cheese factory were plated on M17-agar (Oxoid, Hampshire, England) with 0.5% lactose directly or after 24 hour enrichment on M17-broth with 0.5% lactose (LM17). Single colonies were grown overnight in a microtiterplate with 250 µl LM17. From the grown cultures, 100 µl was used for DNA isolation and to the remainder 50 µl glycerol (85%) was added and the mixture stored at -20°C. To 100 µl cell culture, 50 µl lysozyme solution (30 mM Tris-HCl pH 8.0, 3 mM $MgCl_2$, 25% saccharose and 2 mg/ml lysozyme) was added and incubated for 30 minutes at 37°C. The cells were lysed by the addition of 100 µl 0.5 M NaOH with 1% sodium dodecyl sulfate. The DNA was fixed to a nylon-membrane filter (Gene Screen Plus, Dupont, Boston, Mass.) by blotting 100 µl of the lysed cell suspension using a dot-blot manifold (Minifold; Schleicher & Schuell, Inc., Keene, N.H.).

These filters were hybridized with a general *Lactococcus* probe, which was the amplified V1 and V2 region of the 16S rRNA (primer 1: position 41-60 and primer 2: position 338-358, corresponding to the *Escherichia coli* numbering of 16S rRNA (Neefs et al. 1990)). Prehybridization and hybridization were performed in 0.5 M sodium phosphate buffer (pH 7.2) containing 3% sodium dodecyl sulfate and 1% bovine serum albumin. After 30 min of prehybridization at 65°C, the probe, which had been labelled with [α-^{32}P]ATP by nick-translation (Sambrook et al. 1989), was added and the hybridization was continued overnight. The blots were washed with 0.03 M NaCl-0.003 M sodium citrate with 1% sodium dodecyl sulfate at 65°C until a clear signal was found.

After analysis of the hybridization with the general probe, positive cultures were selected and plated on a LM17-agar. From the grown colonies, the DNA was isolated and the V1 region was amplified using PCR and the amplified DNA was blotted on a nylon-membrane filter using a dot-blot manifold and hybridized with the species specific probes according to Klijn et al. (1991)..

Results and Discussion

DNA from single colonies obtained from environmental samples was first screened with the general *Lactococcus* probe. From the isolates that were positive, 85% could subsequently be identified with species specific probes. After sequence analysis (unpublished data) it appeared that the other 15% consisted of closely related streptococci or as yet unknown species.

The results shown in Table 1 indicate that various *Lactococcus* species can be found in the environment of cattle. By comparing the samples taken on the dairy farm, where the milk of the cows was used for the production of cheese, and the samples taken from the bull raising farm showed that the presence of milk is not a prerequisite for the establishment of *Lactococcus* species.

Table 1. *Lactococcus* species isolated from environmental samples from a dairy and a bull farm.

Sample	*L. lactis* subsp. *lactis*	*cremoris*	*L.garvieae*	*L.raffinolactis*
raw milk	+	+	+	+
milking machines	+	+		
cows udder			+	+
saliva[1]	+	+		+
skin[1]	+		+	+
silage[1]	+			
grass[1]		+		+
soil[1]	+		+	+

[1] = similar results were obtained with samples from the dairy and bull farm
+ = strain(s) detected with specific 16S rRNA probe

To study the survival of *Lactococcus* species in the waste-flow of a cheese factory samples were taken from cheese-whey, wastewater (from rinsing equipment etc.) and subsequent points in the waste-flow of cheese production. *Lactococcus* species were detected in almost all samples. Besides *L. lactis* subsp. also *L. garvieae* and *L. raffinolactis* were detected, mostly downstream of the production site. This suggest that the production of cheese results in a release of living lactococci in the environment.

To test the survival of *Lactococcus* species in the human intestinal tract, stool-sample were screened for the presence of lactococci. Until now we were unable to detect *Lactococcus* species in human or animal faecal samples. To increase the sensitivity of the detection an extraction of DNA from stool samples, which can be used directly in a PCR-amplification is now under investigation.

The results of this study show that the combination of the use of specific DNA probes with PCR-amplification allow the reliable screening of many isolates in a short time period.

Using these techniques, we were able to demonstrate that *Lactococcus* species are generally present in the environment of cattle and that they are a part of the natural bacterial flora in soil and vegetation.

The survival and growth of *Lactococcus* species outside the dairy environment results in an additional factor involved in assessing the biosafety of the use of genetically modified lactococci in the production of fermented food products.

References

Barry T, Powell R, Gannon F (1990). A general method to generate DNA probes for micro-organisms. Bio/Technology 8:233-236.

Klijn N, Weerkamp AH, de Vos WM (1991). Identification of mesophilic lactic acid bacteria by using polymerase chain reaction-amplified variable regions of 16S rRNA and specific DNA probes. Appl Environ Microbiol 57:3390-3393.

Matthews JA, Kricka JL (1988). Analytical strategies for the use of DNA probes. Anal Biochem 169:1-25.

Sambrook J, Fritsch EF, Maniatis T (1989). Molecular Cloning. A laboratory manual. Second edition 10.6-10.12.

Neefs J, van de Peer Y, Hendriks L, de Wachter R (1990). Compilation of small ribosomal subunit RNA sequences. Nucleic Acids Res 18(Suppl.):2237-2317.

Saiki RK, Gelfand DH, Stoffel S, Scharf SJ, Higuchi RG, Horn GT, Mullis KB, Erlich HA (1988). Primer-directed enzymatic amplification of DNA with a thermostable DNA polymerase. Science 239:487-491.

Sandine WE, Radich PC, Elliker PR (1972). Ecology of the lactic streptococci. A review. J.Milk Food Technol 35:177-184.

Schleifer KH, Kraus J, Dvorak C, Kilpper-Bälz R, Collins MD, Fischer W (1985). Transfer of *Streptococcus lactis* and related streptococci to the genus *Lactococcus* gen.nov. System Appl Microbiol 6:183-195.

Schultz JE, Breznak JA (1978). Heterotrophic bacteria present in hindgut of wood-eating termites [*Reticulitermes flavipes* (Kollar)]. Appl Environ Microbiol 35:930-936.

Williams AM, Fryer JC, Collins MD (1990). *Lactococcus piscium* sp.nov., a new *Lactococcus* species from salmonid fish. FEMS Microbiol Lett 68:109-114.

Subject Index

Printing: Mercedesdruck, Berlin
Binding: Buchbinderei Lüderitz & Bauer, Berlin